PROLEGOMENON

Of all species, man is especially subject to gravitational stresses. These are a daily event on assuming the upright position from reclining or lying. The resultant diminution in central blood volume, due to the displacement of blood to the venous capacitance system of the lower body, demands complex adjustments in the cardiovascular system to offset the decrease in cardiac filling pressure. Such changes are necessary to sustain arterial blood pressure at an appropriate level so that there is adequate perfusion of the vital organs. These adjustments must compensate for both a quick and sustained orthostatic stress. The rapid adaptations depend primarily on the cardiovascular reflexes, including the arterial and cardiopulmonary mechanoreceptors and vasomotor centers in the brain, with resultant changes in autonomic outflow to the heart and the arterial and venous capacitance vessels. Also involved are local venoarteriolar reflexes in the lower limbs and myogenic responses of arteriolar smooth muscle to the increased transmural pressure. Later, the plasma concentration of certain vasoactive hormones, including catechol-amines, angiotensin II, and vasopressin are increased. Activation of the skeletal muscle pump in the legs forces venous blood centrally and the venous valves prevent its reflux. This, together with the reflex venoconstriction in the splanchnic bed, enhances the cardiac filling pressure and hence the stroke volume.

With prolonged standing, the increase in capillary pressure in the dependent parts results in an increase in the interstitial volume and a decrease in plasma volume. This loss is attenuated by autoregulatory adjustments in pre- and postcapillary resistance. At this time, the regulatory mechanisms controlling the total blood volume, including antidiuretic hormone and the atrial natriuretic peptides, assume increasing importance. If the subject is undergoing a severe heat stress, this adds an additional burden on the circulatory system due to the large increase in blood flow to the skin to maintain an appropriate core temperature. This demands even more opposing vasoconstriction of arterial vessels and veins, particularly in the splanchnic bed and the kidneys, to maintain an adequate arterial blood pressure.

The exposure to zero gravity in space flight has offered new challenges and provides fresh opportunities to gain understanding of the cardiovascular responses to gravitational stress. Here, by contrast to standing, the circulation has to adjust to the sudden increase in blood volume in the upper part of the body when zero gravity is reached.

Understanding these complex changes and interactions in normal healthy adults of both sexes and the changes during normal aging is necessary in order to appreciate and to treat effectively the damage caused by disease. Vasovagal syncope, which can occur in susceptible individuals with prolonged standing or emotional upset, is attributed primarily to reflex vasodilatation in the skeletal muscles, but requires further investigation to understand the mechanisms involved. Postural hypotension, due to chronic autonomic dysfunction, is a disabling disease. It can be caused by either central or local disturbances in the function of the sympathetic nervous system. The hypotension can also occur after meals, due to failure of compensatory mechanisms to offset the vasodilatation in the splanchnic bed. Diabetic neuropathy also can affect the autonomic nerves and lead to postural hypotension.

No one is better qualified to edit this review volume of the circulatory responses to the upright position than Dr. James Smith. This has been his major interest for many years, during which he has been a prime contributor in the area of hemodynamic response to postural stress, including the rapid changes, the effect of age, and the use of impedance cardiography and lower body negative pressure, which has provided new information. He has assembled a group of younger colleagues, some of whom have trained with him, to provide an excellent overview of current knowledge in this complex field. This, together with the key references, will serve as a challenge and a platform for all concerned with the effect of gravity on the circulation. It will stimulate students, clinicians, and scientists to seek new knowledge relevant to our understanding of normal function and to search for the causes and possible remedies of postural disturbances.

John T. Shepherd, M.D., D.Sc.
Mayo Clinic and Mayo Foundation

PREFACE

There have been several previous reviews of the physiological response to the upright posture. Chief among those describing the circulatory effects were the publications of Hellenbrandt and Franseen in 1943, Gauer and Thron in 1965, and Blomqvist and Stone in 1982. However, the rapid growth of information in the last 10 to 15 years and the increasing realization of the influence of individual and environmental factors have suggested another, broader-based analysis.

Our objective has been to prepare a one-volume review of the general circulatory responses of the human to the upright posture, and to examine how these responses are affected by physical fitness, weightlessness, age, sex, and orthostatic hypotension. Our volume is concerned not only with changes induced by the upright position but also those due to lower body negative pressure — a procedure which has been used extensively to simulate postural stress.

Notes on circulatory methods have been included only to the extent that the contributors believed them relevant to the particular chapter. The volume is primarily concerned with the human response to orthostasis, so that animal work has been introduced only occasionally to further our basic understanding of the mechanisms involved. All of the contributors have had extensive personal involvement in the investigation of postural tolerance and have been encouraged to incorporate relevant portions of their own studies. Authors have also been asked to summarize their areas and to make their own recommendations of the more fruitful directions for future research. We believe the volume may be of interest to circulatory physiologists and cardiologists as well as those concerned with the biological effects of exercise, physical fitness, aging, space travel, and environmental physiology.

In some chapters — especially Chapter 6 — reference has been made to the pathophysiological significance of deviations from the usual circulatory response to the upright posture. There is a growing literature on the use of postural change as a diagnostic and prognostic aid in the assessment of cardiovascular disease. Because of space restrictions we have not made a systematic review of this topic in this volume. However, it seems likely that with the increasing availability of noninvasive methods, particulary for cardiac output, such functional tests will come into more common use in clinical medicine.

I would like to extend my thanks to the contributors, who have been most cooperative, and what is even rarer, very prompt in submitting their manuscripts. Through frequent consultation, we have tried to keep duplication to a minimum. I also wish to express my special appreciation to our consultants, especially Dr. Carl F. Rothe of the Indiana University School of Medicine; Dr. G. Wyckliffe Hoffler of the Kennedy Space Center, Florida; Dr. Brent M. Egan, Director of the Hypertension Section, Medical College of Wisconsin; Dr. Sheldon Sheps of the Hypertension Division, Mayo Clinic, Rochester, Minnesota; and Dr. Elizabeth Turpin of the Ferris State University, Big Rapids Michigan for reviewing various chapters and providing useful and important suggestions. We are particularly grateful to Dr. John T. Shepherd for agreeing to lend the benefit of his considerable experience in cardiovascular research to the introduction and thereby adding valuable perspective to the review. We also extend our thanks to CRC Press for their cooperation, particularly to Marsha Baker, Executive Editor; Janice Morey, Coordinating Editor; and all others who contributed to the production of the book, including Kathi Unger, Rich Ruggieri, and Sharon Smith.

James J. Smith

EDITOR

James J. Smith, M.D., Ph.D., is Professor of Physiology and Medicine at the Medical College of Wisconsin and Director of the Human Performance Laboratory at the Zablocki Veterans Administration Center in Milwaukee, Wisconsin.

Following undergraduate training at the University of Minnesota, he received his M.D. from St. Louis University School of Medicine and his Ph.D. in physiology from Northwestern University. He was on the staff of the Aeromedical Research Lab, Wright Field, Dayton, Ohio, and Director of Research and commanding officer of the U.S. Air Force Medical Research Establishment, High Wycombe, England, where his work was primarily concerned with physiological responses to high altitude flying. He was on the physiology faculties of Loyola University School of Medicine and the George Washington University School of Medicine before becoming Professor and Chairman of the Department of Physiology at the Medical College of Wisconsin. In 1978, he became Professor of Physiology and Medicine and Director of the Human Performance Lab at the VA Medical Center.

Dr. Smith was Fulbright Research Professor at the Institute of Physiology and the Max Planck Institute, Heidelberg, Germany from 1959 to 1960. He is a member of Sigma Xi, Society for Experimental Biology, American Physiological Society, American Heart Association, and the Gerontological Society of America. He has been the recipient of many research and training grants from the National Institute of Health and other agencies; is author of more than 100 papers in the field of circulatory shock, exercise and nonexercise stress; and author or co-author of three books. His current research interests are the effects of physical fitness, aging, and other factors on the human response to circulatory stress.

CONTRIBUTORS

Jill A. Barney, M.S.
Technical Director
Human Performance Lab
Zablocki VA Medical Center
Milwaukee, Wisconsin

John B. Charles, Ph.D.
Head, Cardiovascular Laboratory
Space Biomedical Research Institute
NASA Johnson Space Center
Houston, Texas

Thomas J. Ebert, M.D., Ph.D.
Associate Professor of Anesthesiology and
  Physiology
Medical College of Wisconsin
Milwaukee, Wisconsin, and
Staff Anesthesiologist
Zablocki VA Medical Center
Milwaukee, Wisconsin

Mary Anne Bassett Frey, Ph.D.
USRA Visiting Scientist
Space Biomedical Research Institute
NASA Johnson Space Center
Houston, Texas, and
Adjunct Professor
Department of Community Health
Wright State University School of Medicine
Dayton, Ohio

Diana E. Houston, Ph.D.
Senior Research Scientist
Krug Life Sciences
NASA Johnson Space Center
Houston, Texas

Mahendr S. Kochar, M.D., M.S.
Professor of Medicine and Pharmacology and
Associate Dean for Graduate Medical
  Education
Medical College of Wisconsin
Milwaukee, Wisconsin, and
Associate Chief of Staff for Education and
Chief of Hypertension
Zablocki VA Medical Center
Milwaukee, Wisconsin

Carol J. M. Porth, Ph.D.
Professor, School of Nursing
University of Wisconsin-Milwaukee
Milwaukee, Wisconsin, and
Adjunct Assistant Professor of Physiology
Medical College of Wisconsin
Milwaukee, Wisconsin

James J. Smith, M.D., Ph.D.
Professor of Physiology and Medicine
Medical College of Wisconsin
Milwaukee, Wisconsin, and
Director, Human Performance Laboratory
Zablocki VA Medical Center
Milwaukee, Wisconsin

TABLE OF CONTENTS

Chapter 1

GENERAL RESPONSE TO ORTHOSTATIC STRESS

James J. Smith and Thomas J. Ebert

TABLE OF CONTENTS

I. THE UPRIGHT POSTURE IN THE HUMAN — GENERAL

Bipedal hominids (two-legged primates including humans) lived in east and south Africa at least four million years ago. From that time until the emergence of *Homo sapiens*, about 250,000 years ago, many significant physical changes took place. Among them were brain expansion and the modification of the female pelvis to permit the birth of such larger brained babies. However, during the early stages of that evolutionary period, the most critical adaption was the development of bipedal locomotion.[1]

It is theorized that, four to six million years ago, as the environment changed, certain primates, to increase their environmental scope, descended from the tropical rain forests to the grasslands (savannas) to eat berries, roots, and small animals. Amid the tall grasses, an erect posture facilitated the spotting of predators and potential prey. Charles Darwin (*Descent of Man*) wrote that this freeing of the use of arms and hands, partly the cause and partly the result of the erect posture, likely led to the use of clubs and other weapons; this facilitated feeding, which, in turn, resulted in a reduction in the size of jaws and teeth. Though there is some disagreement about the sequence of events in the development of bipedalism, no one seriously questions its evolutionary significance for the human species.[1,2]

Human bipedalism has three main features, i.e., (1) erect standing with straightened knees; (2) bipedal walking, and (3) bipedal running. These capabilities necessitated some significant musculoskeletal changes, e.g., elongation and reorientation of the bones of the thigh, leg, and foot in order to alter the weight-bearing axis; this permitted a "spring" in the step and facilitated the development of strong gluteal and leg muscles.[2] The distinguishing feature of human locomotion is not erectness per se, but prolonged erectness, which became possible through straightening of the lower extremities and certain physiological adaptations. Only humans can stand erect for long periods of time. It was theorized that the mobile, bipedal male, now capable of more energy-efficient travel, foraged more widely for food. With the advent of single birth and the increasing length of dependency of the infant, a division of labor developed. The female, relieved of the necessity of long-range foraging, collected food locally and now had more time to parent and protect the young.[2]

To stand erect and maintain postural stability required the development of complex and exacting physiological mechanisms, mainly neuromuscular and circulatory. From a musculoskeletal standpoint, the upright posture and mobility are achieved through integration of multiple somatic, neuromuscular reflexes, which maintain mechanical stability during all phases of erectness and movement. The task is considerably complicated by the high center of gravity and the very small support base in the human, which is undoubtedly the reason for the inevitable postural sway characteristic of the human.[3]

The circulatory handicap of the erect posture is a result of gravity and is twofold: (1) in the upright, about 10% of total blood volume is almost immediately relocated from the thorax to the dependent regions and thus temporarily denied to the heart for pumping purposes and (2) when erect, the head lies well above the heart, which must now pump blood against the hydrostatic pressure to provide adequate blood flow to protect the critical cerebral tissue from hypoxia.

Coping with these handicaps requires complex interaction of autonomic reflexes and a well-functioning circulation.[4] Unfortunately, with aging and disease, neuromuscular, autonomic, and circulatory defense mechanisms may be weakened; as a consequence, upon standing upright, postural stability, mobility, and even the maintenance of consciousness may be jeopardized. The human response to standing is heavily influenced by physical fitness, weightlessness, age, sex, and certain circulatory disorders, particularly orthostatic hypotension. Because of the importance of these factors, separate chapters of this book are devoted to these topics.

The objectives of this introductory chapter are (1) to describe the physical factors that influence the amount of blood volume dislocation, which is the all-important determinant of the degree of circulatory stress, (2) to document the circulatory response to postural stress and the mechanisms involved, (3) to contrast the reactions to head-up posture with those of lower body negative pressure (LBNP), a technique widely used to simulate the head-up position, and (4) to discuss certain factors that cause wide variability in postural response, such as intravascular instrumentation, ambient temperature, and emotional stress. There are several excellent previous reviews of the circulatory dynamics of postural change;[4-8] we have not repeated this material but have extracted those portions that are particularly relevant to current problems. While *orthostatic* in the strict sense of the word, refers to the upright posture, which will be the primary emphasis in this review, the term *orthostatic* will also be used in the broader sense to include LBNP.

II. PRESSURE-VOLUME-FLOW RELATIONS IN THE STANDING POSITION

A. GRAVITY EFFECT

In the standing position, gravity will have an important influence on vascular volume and pressures in various parts of the body. The pressures in the lower body are increased and in the upper body decreased depending on the vertical distances from the heart. Upon quiet standing, the foot veins and arteries are about 110 to 120 cm below the heart, thus adding about 80 to 88 mmHg pressure to the blood column at that point.

It is important to note that although the pressures inside the dependent vessels are increased, the driving pressure, i.e., the impetus forcing blood from the arterial side to the venous side and back toward the heart, is unchanged, i.e., it is unaffected by gravity because the increases in pressure on the arterial and venous sides are the same. The effective driving pressure at any point propelling blood toward the heart will, therefore, be the gauge pressure (that relative to the atmosphere) minus the gravity pressure. To estimate this driving pressure, the intravascular gauge pressure may be "corrected" by adding or subtracting the vertical height of the blood column with reference to the heart. However, the gravity (hydrostatic) pressure will be effective laterally, i.e., across the vessel wall, and, in the upright position, will increase transmural pressure and the transudation of fluid from the capillaries.

When posture is not fully upright but only a fraction thereof, the gravitational effect will correspond to the gravity vector exerted. In some experimental postural studies, the subject does not free stand, but is passively rotated to the upright position on a tilt table with the fulcrum near the center of the body mass. In this manner, the subject may be tilted to various degrees of head-up (or head-down) tilt. As a result, the gravity component, and therefore the degree of stress, is a function of the sine of the angle and not the angle itself. Thus, 30° head-up tilt represents 0.5 G, and 70° tilt (the most commonly used angle) induces 94% of the full gravity vector. The physiological responses to free standing, head-up tilt, and LBNP are similar but not identical and are contrasted in Section IX below.

Since, in the erect posture, vascular pressures in the dependent parts of the body rise and those in the upper parts fall, there must exist a neutral zone at which intravascular pressures do not change with posture. The exact location of this level, termed the *hydrostatic indifferent point* (HIP), is not certain.[4,5] The HIP varies for the arterial and venous circulations, and also varies with arterial tone and the degree of postural stress.[4]

Earlier studies by Gauer[4] indicated that in the normal human, the venous HIP is located about 5 to 8 cm below the diaphragm. This infradiaphragmatic location is significant, since it implies that pressures in the heart would decrease during head-up posture and increase in head-down posture, which has been shown experimentally.[4] Using a unique noninvasive method, Kirsch and von Ameln have reported the venous HIP in humans to be about 10 to 15 cm below the heart;[9] these authors believe that this more inferior location, near the postcapillary area of the liver, protects the portal bed from large changes in venous pressure and is physiologically advantageous to the hepatic circulation.

B. CIRCULATION IN THE LOWER BODY
1. Venous Distensibility and Dependent Pooling

The caudal shift of blood occurs in two phases — a large, immediate shift within the first 1 to 3 min, and a second, more gradual one over the next 20 to 30 min; the latter is due to increased capillary diffusion into interstitial tissues of the extremities. The degree of initial volume displacement will depend primarily on the degree of pressure change and the relative distensibilities of the lower vascular segments affected. Following are a few general principles governing dependent pooling.

Distensibility of a vessel may be defined as the percent volume change that accompanies a unit pressure change. Since the distensibility of veins is about six to eight times that of arteries at their respective physiological pressures,[10] fluids dislocated by intravascular pressure changes will migrate primarily to the venous system. There is, therefore, a rapid dependent venous pooling of blood into the lower back, buttocks, pelvis, and the lower extremities.[4,5] The low-pressure system of the circulation, i.e., the systemic venous system to the left ventricle, has been likened by some investigators to a large, passive volume container.[4,8] Figure 1A represents the volume of the low-pressure system in the recumbent position and Figure 1B that in the upright. Head-out water immersion (HOWI) of the upright body in water to the diaphragm, i.e., to about the level of the HIP, will in effect restore the hydrostatic vascular values to about that of the recumbent position (Figure 1A). Immersion to levels above HIP (e.g., to the neck) will proportionally further increase thoracic blood volume (TBV) and central venous pressure (CVP) (Figure 1C), so that CVP will average 10 to 13 mmHg above the supine level.[5] HOWI, which is sometimes used to simulate the weightless state, is discussed further in Chapter 3. In patients with pulmonary vascular congestion, the increased thoracic vascular volume (as in HOWI) encroaches on lung space and limits pulmonary compliance, so that dyspnea, or difficult breathing, may result. While the passive elastic container analogy implies little or no active peripheral venous contraction, it should be emphasized that in other stresses, e.g., hemorrhage or temperature change, the veins play an active vasoconstrictor role[6,8,11] and are likely to be active during HOWI.

As described in the preceding, the upright position combines a gravity force and a highly distensible venous system, which results in a high-volume, high-pressure lower limb vascular bed. Are there any local extravascular tissue elements that might limit or counteract this gravity effect? The main tissues supporting and surrounding the lower extremity vasculature are (1) the dermis and interstitial connecting tissue and (2) the skeletal muscle. The former plays a relatively passive role, the latter an active one.

2. Dermis and Interstitial Connective Tissue

The connective tissue elements and the surrounding dermis will influence the vascular transmural pressure depending on their composite stiffness. A striking example of this is the

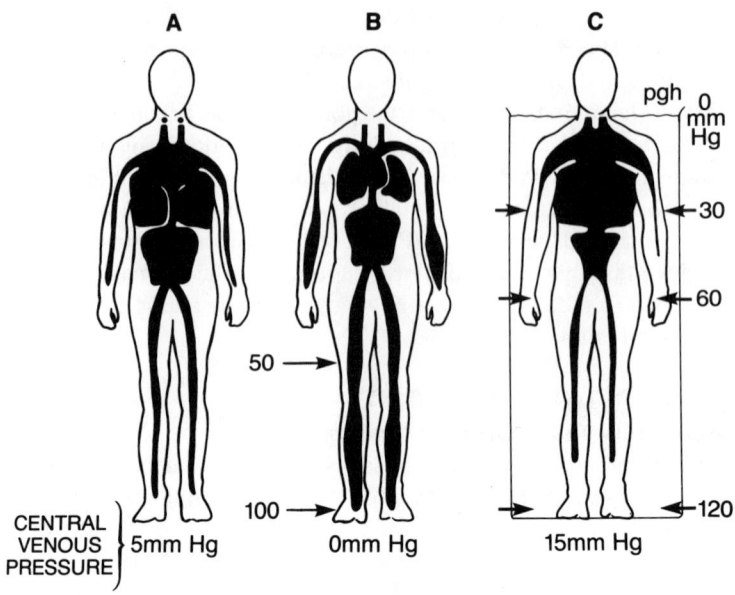

FIGURE 1.   Effect of upright posture and water immersion on blood volume distribution, intravascular pressures (mmHg) in the lower body, and central venous pressure (mmHg). (A) Normal central blood volume and central venous pressure in the supine position are both reduced in the upright position; (B) through footward migration of about 8 to 10 ml/kg of blood (500 to 700 ml in normal adult). Immersion of the upright body to the level of the diaphragm restores volume and pressure relations to approximately those of recumbent position (A). Immersion of the upright body to the neck produces abnormally high central blood volume and pressure (C). (From Rowell, L. D., *Human Circulation During Physical Stress,* Oxford University Press, London, 1986. With permission.)

giraffe whose tough and closely enveloping hide protects its legs against the enormous intravascular pressure.[4] In the human, this situation does not prevail and interstitial fluid in the legs increases during standing as a result of increased capillary diffusion.

3. Skeletal Muscle Pump of Lower Extremities

Skeletal muscle plays an important adjunctive role in promoting venous return so that it is sometimes referred to as a pump or a "second heart".[8] Both voluntary and involuntary contraction of the muscles of the lower part of the body help to milk the venous blood centralwards, and there is little question of its critical role in the prevention of circulatory collapse in prolonged standing.[4,5,7,8,10,13,14] This subject is further discussed in Section VI below.

C. CIRCULATION IN THE MIDDLE AND UPPER BODY
1. Influence of Peritoneal and Pleural Membranes

Early investigators reported that in the abdominal cavity there is, from the diaphragm to the pelvic floor, a linearly increasing extravascular pressure gradient matching that in the inferior vena cava. Thus, transmural pressure of the abdominal blood vessels alters very little because the peritoneal fluid and abdominal viscera represent a water-filled jacket that neutralizes the gravity effect in a manner similar to that of external water immersion of the body.[4,5] Avasthey and Wood showed that in the vertical position pericardial and pleural pressures also minimize gravitational forces in the chest via the water-jacket effect, in a manner similar to that of the intraperitoneal and cerebrospinal fluids.[15]

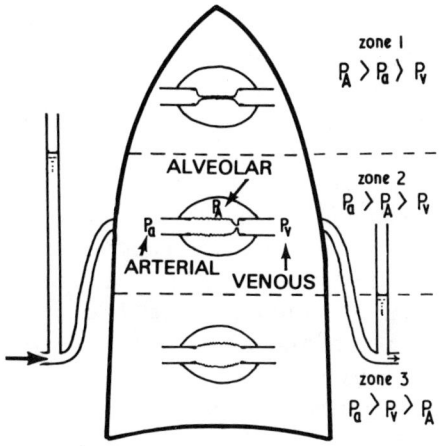

FIGURE 2. Effect of upright position on pulmonary capillary flow. In zone 1 above heart level, the extramural pulmonary alveolar pressure (P_A) may be greater than the pulmonary capillary pressure and some capillaries may collapse. In zone 2, flow may be intermittent. In zone 3 below the heart, with increased hydrostatic pressure, the capillaries distend and flow is increased. (From West, J. B., Dollert, C. T., and Naimark, A., *J. Appl. Physiol.,* 19, 713, 1964. With permission.)

2. Pulmonary Circulation

In spite of the negative influence of hydrostatic pressures on vessels above the heart, perfusion pressures remain positive to all supracardiac tissues because of the level of arterial pressures. However, the gravity effect does play a role in circulation to the more superiorly placed organs and represents a potential hazard, particularly to venous flow. The decrease of thoracic blood volume (TBV) and central venous pressure (CVP) reduces the pulmonary arterial pressure from about 20/7 to about 15/6 mmHg.[4] There is also an increased pulmonary ventilation, a decrease in functional residual capacity, and hypocapnea[13,55,56] with the reduction in pulmonary blood flow at the lung apex, as well as alteration of the ventilation-perfusion ratio.[16] In the upright position, there is less blood flow to the lung apex and progressively increasing flow to the base, as illustrated in Figure 2.

This occurs because in the upright there may be a hydrostatic pressure difference of 20 to 25 mmHg between the apex and base of the lung. Since the normal pulmonary arterial pressure is only about 15 mmHg, the pulmonary intracapillary pressure may be lower than the alveolar pressure. Since the pulmonary capillaries (unlike the pulmonary arterial and venous vessels) are influenced mainly by alveolar pressures, the vessels of the apex may collapse, while at the base the capillaries will be distended. As a result, the ventilation-perfusion (VP) ratio is high at the lung apex and low at the base.[16] In the normal subject, this does not greatly disturb the net VP ratio and there is no serious handicap to gas exchange; however, in case of a disturbance of either ventilation or perfusion, gas exchange may be seriously handicapped. The effects of LBNP on pulmonary physiology are described in Section VII.

3. Cerebral Circulation

The tendency toward a significant negative pressure in the cranium during head-up posture might be expected to jeopardize blood flow in the circle of Willis and endanger oxygen supply to the brain. Fortunately, as many studies have shown, cerebral blood flow remains remarkably constant under most physiological conditions.[4] This is because the central nervous system (CNS) vascular bed is better protected against hydrostatic pressure changes than any other part of the body, by virtue of certain anatomical features and particularly through its relationship with the cerebrovascular fluid. The intravenous and extravascular (essentially cerebrospinal fluid) pressures are balanced at all points in the cranium and spine.[5] In the upright position, the cerebrospinal and venous pressures in the skull are both negative, and in the lower spine both are positive. Thus, intravascular driving pressure is maintained, venous collapse is prevented in the skull, and venous distension is prevented in the lower spine. In addition, several other protective mechanisms exist, e.g., the baroreflexes, cerebral autoregulation, the CNS ischemic reflex, and specialized flow responses to P_{CO_2} and P_{O_2}.[10,17,18] These factors all facilitate cerebral arterial flow. Because of the heavy dependence of brain tissue on aerobic metabolism, there is a tight coupling between metabolism and flow, but the metabolite responsible for this coupling is still uncertain.[18]

A further protective feature of cerebrovascular architecture is the firm attachment of the superficial venous intradural sinuses to the inner surface of the skull. This anatomical fixation further prevents venous collapse, in case these veins are subject to excessive negative intramural pressure on standing. In spite of all these safeguards, there is a decrease of about 15 to 20% in cerebral blood flow during head-up tilt in the normal adult, with an accompanying decrease in perfusion pressure to levels of about 55 to 84 mmHg.[19] This change is readily tolerable in the healthy subject, but in certain pathological situations it can threaten normal cerebral function and even consciousness, (Section VIII below).

III. METHODS FOR THE STUDY OF HEMODYNAMIC CHANGES DURING ORTHOSTASIS

A. CARDIAC OUTPUT

Because interpretation of research results is closely linked to the methods used to obtain them, we believe a brief discussion of methodology is relevant. The most commonly used cardiac output methods in clinical medicine and research are variations of the dye-dilution or thermodilution methods. Repeated determination of cardiac output using the Fick and dye-dilution methods usually agree within a range of $\pm15\%$.[20] Aside from invasiveness, their primary disadvantage is time averaging, i.e., the output estimate depends on dispersion and the sensing time of an indicator. Indirect noninvasive methods based on gas exchange or CO_2 rebreathing are available[21] and are used particularly in exercise. The necessity for a face mask is, however, disadvantageous, particularly with patients. In order to follow cardiac output in a hemodynamically changing situation, e.g., during stress, a noninvasive stroke volume method is preferable. In recent years two procedures of this type have been developed, i.e., pulsed Doppler echo and impedance cardiography. Loeppky et al. have reported stroke volume and cardiac output values with the echo method in supine and erect exercise that compared favorably with previously reported values; they concluded that this was a valid noninvasive technique for studying transient and steady-state situations.[22]

Geddes and Baker have reviewed both the theoretical and practical aspects of impedance cardiography and its application to biology and cardiovascular medicine.[23] Currently the most widely used impedance technique is the tetrapolar system developed by Kubicek and his colleagues,[24] which, in the absence of left-to-right shunts and valvular insufficiency, provides reliable estimates of stroke volume and cardiac output.[25-30] The impedance method has not been widely accepted, partly because of theoretical limitations of the Kubicek equation and partly because of the necessity for breath hold during the determination. However, recent developments with ensemble averaging have largely obviated the latter problem.[28] Miles and Gotshall, in a recent review, have reported that impedance cardiac outputs have a high intrasubject reproducibility and agree within 15% of values obtained by the more traditional methods, which are themselves accurate to within 15% when compared with each other.[30]

B. THORACIC BLOOD VOLUME

Thoracic blood volume (TBV) is a difficult entity to measure, partly because it is an open-ended compartment and partly because of its heterogeneous composition, including, as it does, large central veins, large arterial vessels, the pulmonary circulation, the heart, coronary flow, and other assorted systemic vessels. Theoretically, TBV refers to all blood within the chest. Practically, it can be approximated by measurement of the central blood volume (CBV), which is contained between an injection site into a large central vein and a sampling site on the arterial side. CBV can be determined by adapting the indicator-dilution formula to determine the vascular volume (ml) between injection and sampling sites. Between the pulmonary artery and a systemic artery, Yu reported a volume of about 600 to 610 ml/m^2 in human subjects.[31] Using a similar method, Marshall and Shepherd found mean CBV values of 1034 ml/m^2 between the superior vena cava and the radial artery at supine rest and 722 ml/m^2 while standing, i.e., a loss of about 30% of CBV in the erect position.[32] Although the dye-dilution method is useful, the openendedness of the thoracic vascular compartment, the variations of injection and sampling sites, and dye-mixing problems make the delineation of the CBV somewhat uncertain and comparisons of measurements rather difficult.[33]

Impedance cardiography can also be used to estimate changes in TBV. Because of the marked sensitivity of electrical resistance to liquid content,[23] alterations of baseline transthoracic impedance (ΔZ_0) indicate relative intrathoracic fluid changes in the dog[23,34-36] and humans.[37-39] Ebert et al., found a close parallelism between the decline in central venous pressure (CVP) and the decrease in impedance-determined TBV.[39] Since in the healthy, physiologically stable human subject, CVP and CBV are generally related in a linear fashion, these findings suggest that ΔZ_0 is a reasonable indicator of change in TBV. While the indicator-dilution method provides an absolute measure of blood volume, the impedance method can, at present, provide only relative measures of change. Baseline transthoracic impedance (Z_0) is influenced by factors such as chest girth, electrode placement, and sex of the subject. Interindividual calibration data do not presently exist for conversion of Z_0 to absolute TBVs.

Although they are not identical, in this chapter we will use the terms *thoracic blood volume* and *central blood volume* interchangeably. Since transthoracic impedance theoretically measures all fluid in the chest, we will use TBV as its measured impedance parameter and CBV will be used to refer to the vascular space determined by dilution methods.

C. VOLUME CHANGES IN THE LOWER EXTREMITIES

The fate of the translocated thoracic blood has been studied with calf plethysmography using temperature-compensated, Hg-in-silastic strain gauges.[40] Impedance estimates of leg volume change reportedly show good agreement with strain-gauge determinations.[39]

IV. BLOOD VOLUME DISPLACEMENT DURING HEAD-UP POSTURE

A. TOTAL AND THORACIC BLOOD VOLUME

Thoracic blood volume (TBV) is an integral part of total blood volume. Although there has been no systematic study of the question, it is likely that the two generally parallel each other. Total blood volume is, in most primates, a reasonably linear function of body weight[5] and, in the healthy human adult, averages 70 to 75 ml/kg, i.e., about 2.5 to 3.0 l/m^2 body surface area.[41,42] Individual variations depend mainly on age, climate, and degree of physical activity.[41] There is evidence that total blood volume is importantly influenced by vascular capacity; patients with abnormally large varicosities have oversize CBV with increased stroke volume and cardiac output in the supine position.[4] These findings suggest that total blood volume is adjusted to the vascular capacity of the body during normal daily activity.

Continued orthostatic stimulation appears to be necessary for the maintenance of adequate total blood volume, since forced bedrest, prolonged water immersion, and weightlessness lead to decreased blood volume, and orthostatic hypotension. Since total blood volume is related to orthostasis, so is the excretion of salt and water. Orthostasis, hemorrhage, and positive-pressure breathing all decrease TBV, as well as the excretion of sodium and chloride, but recumbency increases sodium and water excretion. Such effects are mediated, for the most part, by neurohumoral mechanisms and not directly by blood volume alterations; these are further discussed in Section V.

B. BLOOD VOLUME SHIFT — FIRST (FAST) PHASE
1. Thoracic Blood Volume Change
In recumbent man, about 25 to 30% of total blood volume (i.e., about 20 to 24 ml/kg) is "thoracic" or "central" blood volume,[4,31] about 25% is in the liver and splanchnic area,[8] 9% is estimated as pulmonary blood volume,[43] and the remainder is unaccounted for. CBV and lung capacity (as previously mentioned) have a generally reciprocal relationship, e.g., a marked increase in CBV, may encroach on pulmonary ventilatory capacity and induce dyspnea (Figure 1). Upon assumption of the upright posture, the loss of CBV may be likened to a functional hemorrhage into the dependent vascular bed. It has been reported that 26 to 30% of CBV, i.e., about 6 to 8 ml/kg, is displaced from the thorax to the lower portion of the body in the free-standing position.[41]

2. Lower Extremity Volume Change
Efforts have also been made to trace the fate of the displaced thoracic blood. Total blood volume determinations by Asmussen, with and without high-thigh tourniquets, revealed 10.1 ml/kg blood volume in the lower extremities in the recumbent position. After 5 min of 60° head-up tilt, there was an increase of 560 ml (8.0 ml/kg), i.e., a 79% increase in leg volume.[44] Sjostrand, in a "reverse" experiment, measured total blood volume in the standing position with high-thigh tourniquets and then without tourniquets in the recumbent position in five male subjects and estimated that about 11% of the average total blood volume (about 8.25 ml/kg) was shifted from the lower extremities to the rest of the body in the reclining position. About 78% of this blood was taken up by the thorax.[41]

There have been very few systematic studies of the relative degrees of migration of blood to different parts of the lower body during orthostatic stress. In this regard, the experiments of Ludbrook are interesting.[14] He found that with quiet standing, volume changes in the thigh were three times greater and more rapid than in the calf. He thought this occurred because calf veins filled primarily through arterial inflow and thigh veins mainly by reflux from above. The effects of quiet standing on muscular pressures and venous filling are discussed further in Section VI.E below (Skeletal Muscle Pump).

Wolthuis et al. studied blood sequestration during LBNP using an isotope monitoring method over different parts of the body. They found that the isotope density of vascular beds in the pelvis, buttock, and thigh were increased 21, 32, and 35%, respectively, and that of the calf was increased 26%.[45] Because of the greater tissue mass of the thigh and buttock compared with the calf, a significantly greater amount of thoracic blood must have been impounded in the thigh and buttocks than in the calf. It is also of interest that the rate of isotope accumulation was faster in the thigh, pelvis, and buttock, reaching near peak in 2.5 min compared with 5 min for near-peak values in the calf and heel. Although this was LBNP and not head-up posture, there is evidence that from the standpoint of fluid migration the two stresses may be comparable (Section VII).

The studies of Ludbrook and Wolthuis taken together suggest that with orthostatic stress a larger volume of thoracic blood than previously suspected may migrate to the pelvis, buttocks, and thighs, and that some of this fluid may be reflux from the abdominal vena cava and iliac veins (Section VI.E).

C. BLOOD VOLUME SHIFT — SECOND (SLOW) PHASE

While the initial, fast phase of blood volume redistribution described above involves intravascular blood relocation from the thorax to dependent parts, the second, slow phase involves mainly a transcapillary diffusion of fluid. During this second phase, there is no loss of red cells[46] and relatively little loss of plasma proteins, although the latter is not entirely certain.[47] Plasma volume changes are deduced from changes in hematocrit and/or plasma protein.

Thompson et al. reported an average loss of about 11% of the total plasma volume in the standing position; the maximum fluid loss occurred during the first 20 to 30 min.[48] Hagen et al. reported that in their subjects the supine plasma volumes averaged 53.2 ml/kg; by 20 min of free stand, plasma volume had decreased 8.70 ml/kg (16.4%). However, the rate of plasma loss declined sharply with time. At 5 min the total plasma loss was 4.4 ml/kg, for an average loss of 0.9 ml/kg/min during that time; however, from the 5th to the 20th minute the plasma volume loss was only 0.29 ml/kg/min, and from the 20th to the 35th minute, i.e., only 0.01 ml/kg/min, an insignificant amount.[46] On the other hand, Tarazi et al. reported only a 1.7% loss of plasma volume at 5 min and a 4.1% loss at 20 min during a 50° head-up tilt; the reason for the low values in this study is not clear.[49] Since the mass density of blood is a linear function of protein concentration, other investigators measured the time course of blood density, plasma density, and erythrocyte density during head-up tilt experiments in normal young subjects and found that plasma and blood density increased almost linearly with increasing angles of tilt. They found mean decreases of 14% plasma volume at 45 min and 18% at 2 h.[47]

The differences in estimated plasma loss between studies likely reflect variations in experimental procedures and sampling times. Most evidence indicates that plasma volumes decline 11 to 16% in plasma volume during the first 20 min of head-up posture. The rate of plasma loss decreases progressively during the first 30 min of orthostasis. Mellander et al. have suggested that this decrease in plasma diffusion is due to local autoregulatory adjustments in pre- and postcapillary resistance, especially constriction of arteriolar smooth muscles; the resultant sharp decrease in transudation is primarily due to the net decrease in capillary surface area presented to the flowing blood, rather than the decrease in filtration pressure from the constriction of arterioles.[50]

D. RATE OF BLOOD VOLUME SHIFT

Previous studies have indicated that the caudal movement of thoracic blood occurs very rapidly, but there have been few quantitative measurements of the rate of fluid movement. Using a unique ultrasonic miniature plethysmograph to record thickness changes in tissue slices, Kirsch et al. found that within 5 s of head-up tilt there had already begun a rapid decrease in fluid volume in the tissues above the heart and increases in those below the heart.[51] During graded tilt, the fluid changes were positively correlated with the increase in the gravity vectors; bringing the subject back to the supine reversed the effects in almost the same time course.[51]

The rate difference of blood volume migration between phases 1 and 2 is illustrated in Figure 3. Ebert et al. recorded a time plot of TBV decrease (thoracic Z_0 increase) in normal male subjects during a 20 min 70° head-up tilt (Figure 3A).[39] Asmussen et al. followed volume changes with a limb plethysmograph extending to the upper thigh during head-up tilt[52] (Figure 3B). These bimodal curves indicate a rapid blood volume change from the thorax to the lower body in the initial 2 to 3 min, followed by a much slower phase. In Asmussen's study, the combined lower extremities (thigh and leg) increased by about 4.8 ml/kg in the first 5 min. This would mean a rate of increased volume of about 0.9 ml/kg/min for the first 5 min. Over the following 60 min the volume change in the two extremities was only 5.8 ml/kg (0.097 ml/kg/min), i.e., about one tenth of the rapid rate. It is interesting that neither the thoracic-volume nor limb-volume graphs

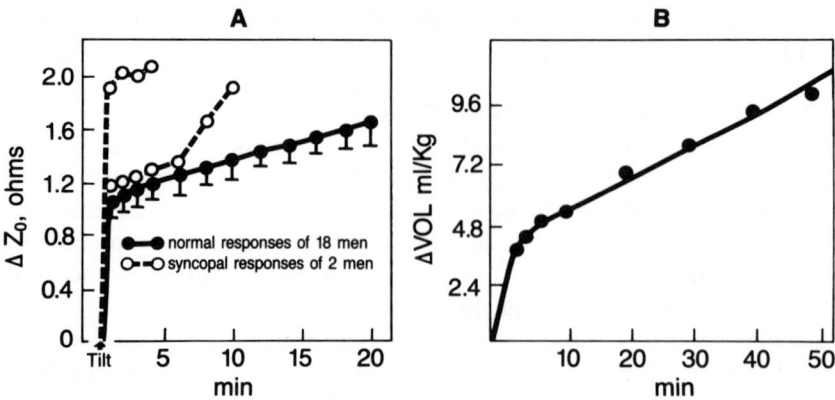

FIGURE 3. Fast and slow phases of blood volume migration in upright position. (A) Decrease in thoracic blood volume (TBV) (increase in transthoracic Z_0) during 20 min 70° head-up tilt in healthy young men. The unusual changes in two subjects who fainted are discussed further in Section VIII. (From Ebert, T. J., Smith, J. J., Barney, J. A., Merrill, D. C., and Smith G. K., *Aviat. Space Environ. Med.*, 57, 49, 1972. With permission.) (B) Increase in volume of both lower extremities (legs and thighs) during head-up tilt. (From Asmussen, E., Christensen, E. H., and Nielsen, M., *Surgery*, 8, 604, 1940. With permission.)

(Figures 3A and 3B) reflect the decreasing transcapillary diffusion reported by investigators who monitored plasma-volume rather than leg-volume changes.[46,47,49] The significance of the TBV changes in the syncopal responses in the two subjects in Figure 3A is discussed in Section VIII below.

From the foregoing, it is possible to make approximations of the rate and degree of blood volume dislocations in the head-up position of healthy young adults. Assuming an average CBV of 22 to 26 ml/kg[4,31] and a 30% decrease of CBV in the upright posture,[32] an average of about 7.2 ml/kg is apparently dislocated from the thorax to the lower body in the first 5 min. An indefinite fraction, probably most of this, gravitates to the legs and thighs (Section VI.E). If about 6%[46,48,49] of an average plasma volume of 52 ml/kg[46] diffuses to interstitial tissue in the first 5 min — most to lower limbs — an estimated additional 3.1 ml/kg of plasma would migrate. This would mean an average of about 10.3 ml/kg increased volume in the lower body in the first 5 min. From the 5th to 20th minute an estimated additional 3.0 ml/kg of plasma filtrate would be sequestered in the lower body, mainly in the lower limbs.

It is clear that this approximation is indeed based on fragmentary information and emphasizes the scarcity of key data on the rate and degree of blood volume dislocation, the best single measure of the degree of postural stress. Certainly this aspect deserves considerably more investigative attention than it has received in the past.

V. COMPENSATORY RESPONSES TO THE HEAD-UP POSTURE

A. NONCIRCULATORY RESPONSES

The complex somatic reflexes necessary to maintain mechanical stability while standing are subject to fatigue. Hellebrandt and Franseen marveled that the large weight-bearing muscle mass could maintain verticality for such long periods without obvious fatigue,[13] but this apparent lack of fatigue may be somewhat illusory. In the experience of fighter pilots in World War II and subjects in the human centrifuge, even a short exposure to G_z (head-to-seat) acceleration results in a considerable decrease in cardiac output and cerebral blood flow with excessive fatigue. In cardiac patients, fatigue is usually ascribable to a reduced cardiac output. Prolonged standing, with a 15 to 20% decrease in cardiac output[53] and cerebral blood flow[19] and a 19% increase in metabolic rate,[54] would undoubtedly take its toll in eventual fatigue. Further diminution of cardiac output from standing in the heat or on a very hard floor (with resultant handicapping of the skeletal muscle pump [SMP]) will accentuate the fatigue.

B. CENTRAL CIRCULATORY RESPONSES

Thus far only the immediate hemodynamic results of the dislocation of thoracic blood have been described without reference to the reactive mechanisms that are called forth to neutralize this dislocation. In this section will be described the compensatory responses — both central and peripheral — that follow rapidly and are the means by which the body copes with the hemodynamic stress. In the subsequent section (VI) will be discussed the mechanisms of these compensatory responses. This review will refer primarily to results obtained in healthy, young male adults about 20 to 35 years of age, as recorded in the first 5 to 10 min of free standing or head-up tilt. By this time the responses are usually stabilized. This population and experimental condition were chosen because the great majority of reported studies were done under these circumstances. In subsequent sections of this chapter will be discussed immediate (first 30 s) and longer term (20 min or longer) responses to the head-up posture. The effects of LBNP, which resemble those of the upright posture, will also be described in a later section.

Previous studies indicate that during head-up tilt decreases in stroke volume ranged from 30 to 45%,[22,53,57,59,62] and the decreases in cardiac output from about 16 to 27%.[22,53,57,59,62] There is invariably an increase in diastolic pressure of about 12 to 17%,[52,53,57,58,62,64] with relatively small increases in mean arterial pressure of about 2 to 10%.[53,57,58,60,65] Systolic pressure is usually unchanged, but a decrease in pulse pressure is practically a universal finding.[52,53,60,61,63] Total vascular resistance increases range from 30 to 65%, but most often are 30 to 40%.[22,53,57,60,63] Heart rate increases vary from 15 to 50%, with most being in the 26 to 43% range.[22,53,57,58,61,63,64,65,67] During head-up tilt, the arteriovenous (A-V) oxygen difference usually increases by 1 to 2 vol%.[4,22] Katkov et al. reported an 18% decrease in coronary sinus blood flow, a 26% decrease in myocardial oxygen consumption, and a 32% increase in coronary vascular resistance.[66] Typical hemodynamic responses during 70° head-up tilt in healthy young men are shown in Figure 4. These changes are usually accompanied by a fall in the right atrial pressure of about 4 to 6 mmHg and a decrease of about 25% in CBV.[8]

C. PERIPHERAL CIRCULATORY RESPONSES

Because of the decrease in cardiac output and the heightened sympathetic vasoconstriction (except in the heart and brain), the flow to all peripheral beds is reduced, particularly to the splanchnic, skeletal muscle, and renal tissues. As estimated by clearance techniques, flow to the kidney declines by about 50 to 60%.[5,7,8] The mechanism of this reduction is difficult to determine because of the time required to measure renal blood flow and the autoregulatory capacity of renal vessels.[7]

Although brain oxygen uptake reportedly changes little in the upright position,[5] cerebral blood flow decreases approximately 10 to 20%.[4,19] Rowell has estimated that this decline is only about half of what it would be without the autoregulatory capability of the cerebral vessels.[8] Because of the sensitivity of cerebral flow to P_{CO_2}, the hypocapnia accompanying the head-up posture may be a factor in reducing cerebral blood flow in orthostasis.[56] In the head-up position, splanchnic blood flow decreases about 40%[8] or about 0.6% for each beat per minute increase in heart rate, an indication of a possible parallelism between heart rate and splanchnic vascular resistance. Rothe has stated that, because of its high capacitance, the splanchnic bed may redistribute as much as 5 ml/kg, i.e., about 350 ml or more, to the general circulation in orthostatic or other types of stress.[6]

In the upright posture, forearm blood flow is decreased 42 to 56% and calculated vascular resistance is increased 60 to 100%.[68] These authors made the interesting observation that with

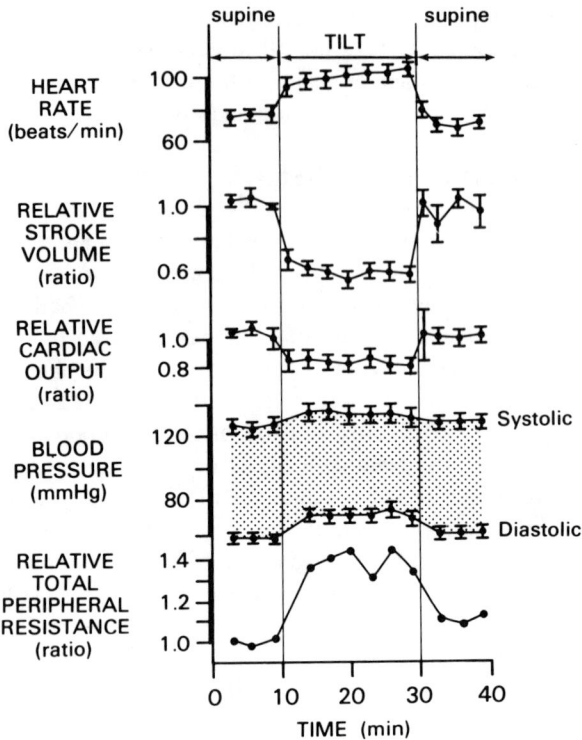

FIGURE 4.   Cardiovascular responses of healthy young men to a 20-min 70° head-up tilt. (From Smith, J. J., Bush, J. E., Wiedmeier, V. T., and Tristani, F. E., *J. Appl. Physiol.*, 29, 133, 1970. With permission.)

a change from supine to sitting, forearm blood flow decreased by 48% but calf blood flow increased by 4%, indicating a marked difference in the responsiveness of these two beds to orthostatic maneuvers.[68] A recent study on static exercise from the same lab further suggests a heterogeneity of forearm and calf response.[69] The reason for this difference is unclear. Previous studies have indicated that with increasing angles of tilt, forearm blood flow decreases[70] and efferent sympathetic nerve activity increases,[71] both in a proportional manner. Similarly, tibial sympathetic nerve activity increases from the lying to the sitting position.[72] A recent study indicates a dissociation between reflex increases in vascular resistance and sympathetic nerve activity in the human,[73] which may account for some of these apparent inconsistencies in vascular responses in the limbs.

   The resting vascular resistance of skin is higher than that of other tissues but, in contrast to that of splanchnic, renal, and forearm vessels, human skin resistance vessels usually show only transitory alteration in tension during the head-up posture. Only when there is a marked decrease in the activity of arterial or cardiopulmonary receptors does a sustained constriction of skin resistance vessels contribute to the systemic vascular resistance.[7] Rowell, however, attributes decreased blood flow in the forearm about equally to skeletal muscle and skin, and concluded there was a sustained decrease in skin blood flow.[8] The resistance vessels of the skin of the foot reportedly have a higher resting tone than skin in the upper extremity, perhaps because foot skin has adapted to the habitual upright posture.[5] Linde and Hjemdahl found a marked increase in the vascular resistance of adipose and skeletal muscle tissue within 2 min after head-up tilt, but shortly thereafter there was vasomotor "escape", i.e., a subsequent decrease in vasoconstriction; however, this study involved some rather strenuous invasive procedures, which may have contributed to the results.[65]

Over the years many investigators have concluded that contraction of the veins in response to postural change is very limited, and Gauer and Thron puzzled over the fact that CBV was not better protected against the caudal sequestration.[4] The lack of venoconstriction in the lower extremity is rather surprising, since peripheral veins are capable of very forceful contractions under other conditions.[5] Rothe states that, in mammals, reflex control of capacitance vessels can act very rapidly to move blood to or from the thorax within an estimated range of about 20% of TBV; however, which autonomic baroreceptor system is responsible for this is not known;[6] this lack of understanding of venous response to hemodynamic stress is due to the fact that venous reactivity has been studied primarily in the limbs and that responses in animals have been too liberally extrapolated to the human.[6] Undoubtedly progress in human venous physiology has also been considerably stymied by the lack of adequate methods.

D. IMMEDIATE CIRCULATORY RESPONSES

Because the immediate (first 30 s) circulatory change upon assumption of the upright posture is an indicator of autonomic neuropathy, there has been increasing interest in this subject, particularly among European investigators. One of the first reports of this type was by von Drischel et al., who studied the early postural responses in 300 normal subjects of both sexes and various ages. Within 1 s after standing, the heart rate rose sharply, peaking to about a 30% increase in 12 to 16 s, then falling abruptly to about a 10% increase level in 20 to 30 s. This was followed by a gradual heart rate rise over the next 10-min period, punctuated by respiratory waves of about 10-s duration. von Drischel described various patterns of this initial "heart rate complex" based on sex and time of day.[74] Brauer and Rossberg compared the initial heart rate complexes of healthy young males to different rates of head-up tilt with that of free standing. They found that with a rapid (1.6-s) 70° head-up tilt and rapid standing (4 s) there were initial heart rate complexes that were similar but more variable than those reported by von Drischel.[75] With slow (30-s) 70° head-up tilt, there was no initial heart rate complex at all; thus the speed of tilting was an important factor in determining the initial heart rate complex.[75]

In a further study by Rossberg and Martinez, rapid (1.7-s) 70° head-up tilt produced similar pronounced heart rate complexes, with peaks about 25% higher and of longer duration if done during expiration than during inspiration.[76] Ewing et al. compared the heart rate complex after fast (5-s) and slow (10- to 15-s) free stand and fast (2- to 3-s) and slow (12-s) head-up tilt. They reported that the initial complexes to slow and fast standing and fast tilt were similar, but in the case of tilt the secondary blood pressure fall at 16 to 20 s was absent; the initial heart rate complex after slow tilt was entirely absent.[77] Autonomic blockade experiments with atropine and propranolol indicated that the initial heart rate complex was due to a quick vagal withdrawal rather than sympathetic stimulation.[77] They reported the standing heart-rate response to be very reproducible in normal individuals; in diabetics these investigators found a flat heart-rate response, due to reflex pathway damage, and believed that this was a valid test for autonomic dysfunction.[77]

Wieling and his colleagues have also studied the characteristics of the initial complex and its clinical application. They found a resemblance between the heart rate complex after free standing and the initial heart rate increase seen during handgrip, and suggested that in both cases the initial heart rate peak and the abrupt decrement were vagal manifestations and were due to the exercise response to standing.[78] Wieling also analyzed several autonomic tests and reported that the heart rate response to standing and to forced breathing were among the most useful for the clinical diagnosis of autonomic disorders.[79,81] As shown in Figure 5, the initial heart rate complex is much reduced in autonomic neuropathy.

Other immediate hemodynamic responses to the head-up posture have been studied to a much lesser extent than heart rate. The blood pressure changes in free stand have been described by Borst et al.[78] and Wieling.[79] Upon standing there is an immediate, very brief arterial pressure increase, then a sharp fall, which reaches its lowest point in about 5 s; this is followed by a rapid

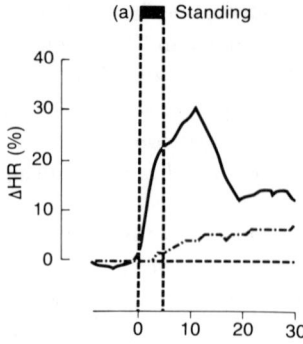

FIGURE 5. "Initial heart rate complex" induced in the first 30 s of standing. Average changes in five healthy adults (average age 45 years) (~) and five age-matched diabetic patients with cardiac vagal neuropathy (---). (From Wieling, W., et al., *Clin. Sci.*, 64, 581, 1983. With permission.)

rise, and sometimes overshoot, which returns the blood pressure to near control levels in about 12 to 18 s. These blood pressure changes resemble those seen at the onset of bicycling.[79]

Smith et al. studied the transient hemodynamic changes to rapid (2.1-s) 70° head-up tilt in male subjects of different ages. In young males, there was an immediate heart rate rise, which peaked in about 6 to 8 s followed by a decline, which reached its low point in about 15 to 22 s, with a subsequent slow rise. There was a progressive fall in stroke volume and TBV, and a temporary short-lived increase in cardiac index, which peaked in about 4 to 5 s and then declined toward and below baseline.[82] The gradualness of the TBV fall compared with the rapidity of the heart rate rise suggested that the initial heart rate complex was not associated with a decreased CVP, but rather that the heart rate was — as previously suggested — more likely due to an exercise response similar to that which occurs with handgrip.[79,80] As described in Chapter 4, middle-aged and older subjects tended to show lesser and more delayed peak heart rate responses and a tendency toward greater decreases in TBV during the first 30 s of 70° head-up tilt.[82] These changes are shown graphically in Figure 2, Chapter 4.

E. NEUROHORMONAL RESPONSES
1. General
The head-up posture, by virtue of its resultant decrease in central venous and arterial baroceptor pressure and reduced perfusion of the hypothalamic osmoreceptors and the renal juxtaglomerular apparatus, causes significant increases in the plasma concentration of a number of hormones, including catecholamines, renin, arginine vasopressin (AVP), and significant reductions in plasma levels of atrial natriuretic factor (ANF).[5,7,8,83] These hormonal changes initiate renal and cardiovascular responses that tend to minimize the loss of body water and promote vasoconstriction, thereby assisting in the maintenance of blood pressure in the upright position. Because of their widespread circulatory effect, these neurohormones have recently become a very active field of physiological investigation. However, because (compared with autonomic neurogenic reflexes) they seem to play a minor role in the early and rapid hemodynamic adjustments to the upright posture, they will be only briefly considered in this chapter. The role of these neurohormones in the diagnosis and treatment of orthostatic hypotension is often critical and is discussed in Chapter 6 of this review.

2. Catecholamines
It has long been known that orthostasis results in activation of the sympathetic nervous system and a measurable rise in plasma norepinephrine (NE) levels.[84] The predominant sources

of circulatory NE in humans are sympathetic nerve endings and, to a much lesser extent, the adrenal medulla. Epinephrine derives primarily from adrenal medullary secretion.[85]

The measurement of peripheral vein plasma levels of NE is a reasonable index of overall sympathetic nervous activity, directed primarily to the vascular smooth muscle, although a considerable amount of NE is rapidly metabolized prior to entering the circulation and only a small fraction of released norepinephrine "spills over" into the venous system.[86]

In the upright position, there is clearly a predominance of neurogenic compared with adrenocortical NE release.[87] There is relatively little change in plasma epinephrine levels with upright tilt until a presyncopal stage is reached.[88,89] There are early measurable increases in plasma NE levels during mild, nonhypotensive head-up tilt (15 to 30°) and low levels of LBNP (10 to 20 mmHg), which unload cardiopulmonary baroceptors and elicit reflex sympathoexcitation.[89,90] Greater degrees of head-up tilt or LBNP unload both arterial and cardiopulmonary baroceptors and elicit more pronounced sympathetic activation, which leads to larger measurable plasma concentrations of NE and, on occasion, small increases in plasma epinephrine levels.[89,90,91] Recent studies suggest that within 1 to 2 min of head-up tilt, plasma levels of NE increase two to threefold,[65,87] which make postural stress the second most potent releaser of NE after upright exercise.[87,92]

3. Renin-Angiotensin-Aldosterone Axis

Activation of the renin-angiotensin-aldosterone (RAA) system occurs gradually over extended periods of orthostatic stress. Relatively brief (10-min) exposure to nonhypotensive levels of LBNP does not appear to elicit significant activation of the RAA axis.[93] However, more sustained (20 to 30 min), nonhypotensive LBNP or head-up-tilt stimuli elicit increases in plasma renin, angiotensin, and aldosterone levels in humans.[89,94-96] These data suggest that the RAA axis is under low-pressure baroceptor control in humans. Moreover, hypotensive orthostatic stress elicits further activation of the RAA system, which suggests that arterial baroceptors may participate in the neural regulation of this hormonal axis as well.[96] The liberation of renin from the kidney leads to elevation of plasma angiotensin II levels, which in turn has several systemic actions, including (1) an increase in vascular resistance by direct augmentation of arteriolar and venular vasoconstriction and (2) a generally enhancing effect on sympathetic activity.[8]

In pathological conditions of autonomic nervous system failure, renin release is augmented in response to head-up tilt compared with healthy control subjects.[97] Since the sympathetic innervation of the kidneys is absent in these pathological conditions, the increase in renin is probably due to decreases in intrarenal vascular pressure. It is conceivable that this heightened activation of the RAA plays an important role in pathological states of autonomic failure.

4. Arginine Vasopressin

Arginine vasopressin (AVP) is one of the most potent vasoconstrictor agents known; however, its participation in the regulation of blood pressure during orthostatic stress is relatively minor in healthy individuals. Plasma levels of AVP can be suppressed by prolonged (2- to 4-h) water immersion.[98] However, moderate increases and decreases in CBV that do not change blood pressure do not alter plasma levels of AVP in humans.[90,99-102] These data suggest that in humans the low-pressure cardiopulmonary baroceptor reflexes do not participate significantly in the regulation of AVP. It is well known that osmotic stimuli can provoke AVP release.[100-102] In addition, AVP may be under high-pressure, arterial baroceptor control; sitting, standing, and high levels of LBNP and head-up tilt can all produce small increases in plasma AVP levels.[89,91,99,103-105]

AVP may play an important role in the recovery of the circulation during orthostatic hypotension and syncope. In these severe situations, plasma levels of AVP rise strikingly to levels 5- to 30-fold above basal levels.[89-91,104,106] These large circulating concentrations of AVP may exert significant clinical effects, as noted by Ebert, who studied the cardiovascular responses to graded infusions of AVP in human volunteers.[107] He demonstrated that AVP can produce significant increases in forearm vascular resistance, total peripheral resistance, and blood pressure. The stimulus for the profound augmentation of plasma AVP during syncope may be hypoperfusion of the pituitary region or altered baroceptor activity, since patients with autonomic insufficiency fail to augment their plasma concentrations of AVP during orthostatic stress.[108] However, another possible mechanism is discussed in Section VIII below.

5. Atrial Natriuretic Factor (ANF)

ANF is a 28-amino acid peptide that is synthesized and released primarily from the cardia atria. The primary stimulus for release is atrial stretch. For example, when cardiac filling pressures are increased by intravascular volume expansion, plasma ANF levels increase.[109,110] It has been hypothesized that this hormone plays a permissive role in promoting the excretion of sodium and water in the kidneys.[111,112] High concentrations of ANF may also inhibit the release of renin and synthesis of aldosterone.[111,112] However, ANF's major effect in physiological concentrations is to reduce intravascular volume via the movement of extracellular fluids to the interstitial space, thereby reducing cardiac filling pressure.[113-115]

Presumably, gradual reductions in circulating levels of ANF that occur with head-up tilt [109,110,116,117] could promote renal retention of sodium and water by direct renal effects or by promoting increases of plasma renin and aldosterone levels. ANF might also reduce the continuous process of filtration of fluids in capillary beds. However, at present, the role of ANF in fluid conservation during head-up tilt appears to be, at best, only minor in magnitude compared with other neurohumoral factors.

VI. MECHANISM OF CIRCULATORY RESPONSE TO ORTHOSTASIS

A. GENERAL

This section will be concerned with how the compensatory reactions to postural changes are brought about, i.e., the mechanisms concerned. The adaptive responses follow immediately after standing. Cardioacceleration begins within 3 to 5 s so that the two processes, i.e., the hemodynamic dislocation of blood and the autonomic regulatory response, merge and proceed simultaneously. Within 10 to 20 s, other autonomic reactions are well under way and by the third to fifth minute the regulatory response is well stabilized.

The main organ systems that bring about the acute adaptation to the head-up posture are (1) the vascular smooth muscle of the arterial and venous beds, (2) the autonomic nervous system, and (3) the skeletal muscle pump (SMP) of the lower extremities. Although all of these are mobilized simultaneously, they have different thresholds of activation and different time patterns of response. However, their degree of response are all tied to the degree of stimulus, which is mainly the amount of blood dislocated in the orthostatic process. This has been amply demonstrated by graded head-up tilt and graded LBNP experiments described in other sections of this chapter.

The vascular smooth muscle adapts directly by contraction or expansion of those vessels whose diameters have been changed.[6] Autoregulatory reactions — particularly of the cerebral and renal vessels — are also important. Folkow effectively demonstrated that during hypotension a reduced stretch of vascular smooth muscle is followed by a relative vasodilation; conversely, increased wall stretch causes the opposite reaction.[118] This effect is apparently a local one and occurs independently of nervous control, but its activation requires a distending pressure of 50 mmHg or more. The phenomenon apparently does not occur in veins.

Other mechanisms play a lesser role. Some investigators have reported postural defects after cerebellar or vestibular defects in animals,[119] but the relevance of these studies to the human is

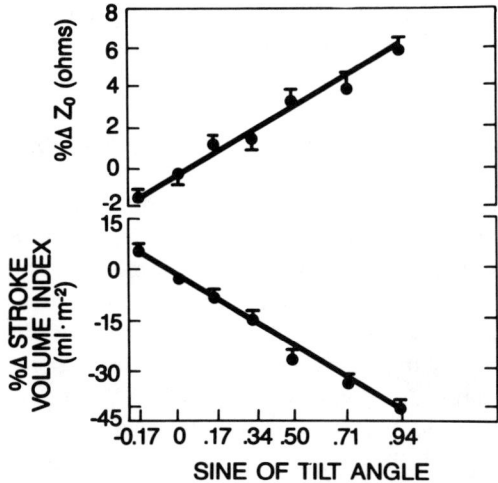

FIGURE 6. Mean percent decrease in TBV (percent increase in transthoracic Z_0) (upper) and mean decrease in stroke volume index (lower) in healthy young men during graded tilt from $-10°$ (g = -0.17) to 70° head-up tilt (g = +0.94). Both of these variables are strongly gravity dependent. The stroke volume changes illustrate the consistency of the Frank-Starling responses through a wide spectrum of different preloads. (From Smith, J. J., Hughes, C. V., Ptacin, M. J., Barney J. A.,Tristani, F. E., and Ebert, T. J., *J. Gerontol.*, 42, 406, 1987. With permission.)

still unproven. Neurohormones, as previously described, are important, particularly in maintaining body fluid balance, but their role in the acute hemodynamic response to orthostasis is probably minor.

B. CARDIOVASCULAR RESPONSES

The immediate consequence of the caudal sequestration of blood is a rapid fall of central venous pressure (CVP) from levels of 4 to 6 mmHg to about 0 to 2 mmHg with an increase in volume and pressure in blood vessels below the heart. In the brief period before autonomic reflexes begin to take effect, there is a passive readjustment of venous and arterial volumes to their altered degrees of distension. With the fall in central venous filling pressure, i.e., decreased preload, there is a rapid decrease in the volume and pressure of all the cardiac chambers, resulting in an immediate fall in stroke volume and cardiac output.

Rothe has emphasized the extreme sensitivity of stroke volume to right atrial pressure; a decrease by 1 cm H_2O of right atrial filling pressure may reduce cardiac output by as much as 50%.[120] However, this sensitivity is progressively decreased if blood volume is decreased, partly because of the lesser venous distension and partly because of the compensatory increases in cardiac rate and contractility. The sensitivity of stroke volume to preload is further demonstrated by the close relation between the decreases in thoracic blood volume (TBV) and stroke volume in graded head-up tilt (Figure 6). Figure 6 illustrates the near-linear decrease in TBV (increase in transthoracic Z_0) and in stroke volume index with increasing gravity increments during head-up tilt in healthy males 20 to 29 years of age.[57]

Coincident with the local volume-pressure alterations in the arteries and veins is the activation of highly important autonomic reflexes. The decrease in pressure at the sites of the cardiopulmonary and arterial baroceptors results in increased heart rate and cardiac contractility, as well as vasoconstriction of peripheral arteries and arterioles of skeletal muscle and splanchnic and renal beds, with a redistribution of venous volumes, especially of the splanchnic bed, into the general circulation.[4,5,11,120,121] A further important consideration in the early stage of the postural response is the absolute necessity of an adequate central venous volume and mean venous filling pressure. It is evident that an increased heart rate and heightened peripheral resistance, while obviously helpful, will be of no avail if there is not adequate venous return. These points have been repeatedly emphasized by early investigators such as Wiggers and Hellebrandt.[13]

Because it is not always practical to study postural cardiodynamics with direct methods, some investigators have monitored cardiac responses in the upright through the timing of cardiac events, using systolic time intervals (STIs). These determinations are usually made by carotid pulse tracings in combination with the ECG. However, because such recordings require patient immobility and cooperation, which are not always attainable, impedance cardiography has also been used. The impedance method does not require patient cooperation and yields STI values within 1 or 2% of conventional methods.[122-126]

Stafford et al. have reported that after 6 min at 90° head-up tilt, healthy young subjects had increases in heart rate averaging 29%, decreases of 32% in left ventricular ejection time (LVET), and decreases of 2% in QS_2 (total electromechanical systole); there were also increases of 24% in the preejection period (PEP) and of 64% in PEP/LVET. These STI changes (with all but PEP/LVET corrected for heart rate) were generally progressive with the sine of the tilt angle.[58] Similar STI changes were reported during head-up tilt by other investigators.[53,61,124-127]

If other factors are held constant, PEP increases with a lessening of ventricular filling or stroke volume, or with an increase in diastolic pressure.[127] LVET is reported to be negatively correlated with heart rate and positively correlated with stroke volume; the quotient PEP/LVET, which is independent of heart rate, is inversely correlated with ventricular ejection fraction and may reflect "ventricular contractility".[124,125,127] In head-up tilt the increased PEP has been interpreted as due not only to the decrease in vascular filling and the increase in aortic diastolic pressure, but also to the slower rise in left ventricular pressure during isovolumic contraction of the ventricle, i.e., a decrease in myocardial fiber length.[61,125,127] The increased PEP/LVET is consistent with, but not necessarily proof of, a diminished cardiac "contractility", i.e., a lesser speed of fiber shortening.[125] STIs have been called a congruent, sensitive, and reliable method for rapid assessment of gravitational effects on the circulatory system, particularly in assessing pathophysiological changes.[61,77,126]

C. AUTONOMIC REFLEXES

Blood migration in healthy young adults usually causes a rapid fall of about 30 to 40% in stroke volume and a 20 to 25% increase in heart rate; the latter cannot compensate for the stroke volume decrease, so there is a 15 to 20% decline in cardiac output. As illustrated in Figure 6 there is a linear relation between the degree of gravity stress (sine of the tilt angle) and the decrease in TBV and, in turn, between both of these variables and the decrease in stroke volume.[57] In the head-up position, total vascular resistance increased about 35 to 40% and diastolic pressure about 10 to 15%, with little or no change in mean pressure. Systolic pressure is usually unchanged but may decline with prolonged orthostasis, especially in older subjects. Pulse pressure falls, due primarily to the rise in diastolic pressure.[4-8,53,56-64] These changes are shown graphically in Figure 4.

Much previous investigation of autonomic mechanisms has been carried out in anesthetized animals; while such studies are indispensable for the study of circulatory mechanisms, recent evidence indicates important differences in the human. Since there are many physiological similarities between LBNP and the head-up posture, the following discussion of mechanisms will include relevant LBNP studies.

1. Arterial Baroreflexes

Two methods are commonly used to study these receptors in human subjects. The pharmacological approach involves the determination of the linear regression of the R-R interval on the systolic pressure following injection of an α-adrenergic agonist (usually phenylephrine); the slope reflects arterial baroreflex sensitivity.[121] The other technique involves direct mechanical stimulation of the carotid baroceptors through suction or pressure inside a cuff encircling the anterior part of the neck. These methods — both of which supply very useful information — have been described and contrasted by Eckberg.[128]

The carotid sinus baroceptors are situated at the base of the skull, where, upon assumption of the upright posture, near-maximum decreases in arterial pressure take place; they are therefore particularly well suited to defend the constancy of the perfusion pressure of the brain. It has been theorized that in the upright position, even in the absence of central arterial pressure change, cardiovascular reflexes may be mobilized by the negative hydrostatic pressure effect (about 15 to 20 mmHg) due to movement of the carotid baroceptors to a more superior position relative to the heart.[4,128] From a hydrostatic standpoint, it seems likely that the aortic baroceptor effect during postural change is likely to be considerably less. Some of the differences noted between the effects of LBNP (usually carried out in the supine position) and postural change (further discussed in Section IX below) may be due to the differential hydrostatic effects, perhaps on cardiopulmonary as well as arterial baroceptors. The primary stimulus to activation of the arterial baroceptors is a change in the transmural pressure at the receptor site. While mean arterial pressure change is often used as an index of possible stimulation of the arterial baroceptors, it is well known that changes in the pulse pressure are particularly effective in stimulating these receptors.[5,8]

There is considerable evidence that, in the human, arterial baroceptors play the major role in regulating cardiac response to stresses such as orthostasis. Pickering et al., using their pharmacological method, reported that reflex bradycardia after phenylephrine was less in the erect than in the supine position;[121] they ascribed this effect either to a centrally mediated inhibition or simply to the result of a higher basal heart rate in the upright position. Eckberg[128] and Robinson et al.[129] suggested that the orthostatic tachycardia was more likely a sympathetic than a parasympathetic effect; it appears probable, however, that the immediate (first 2 to 5 s) cardiac response to the upright position is due mainly to vagal inhibition, with a subsequent combined effect of vagal and sympathetic activation.[77,79]

2. Cardiopulmonary Reflexes

Cardiopulmonary receptors with afferent vagal fibers exert a continuous restraint over sympathetic adrenergic outflow to resistance and capacitance vessels. Decreased pressure at these sites incident to orthostasis, will "unload" the receptors, augment sympathetic vasoconstrictor and venoconstrictor outflow, and also initiate the release of certain neurohormones concerned with fluid retention and balance. One method of delineating the relative roles of cardiopulmonary and arterial baroceptors is the use of a graded LBNP or head-up tilt. Since low-level LBNP (–10 to –15 mmHg) and low-level tilt (10 to 20°) elicit decreases in CVP but little or no tachycardia or arterial pressure change, its effects are presumed to be due to cardiopulmonary receptors.

Such low levels of LBNP produce a significant increase in forearm vascular resistance,[8,130,131] an increase in the resistance in splanchnic vasculature[8] and the kidney,[6-8] and increases in total peripheral resistance.[131] It is of interest that the greater part of forearm vascular resistance at high-level LBNP was already achieved at low levels in young men,[130,131] but not in middle-aged or older men;[131] however, of the total peripheral resistance at high-level head-up tilt, over 70% was achieved at low levels in middle-aged and older (but not young) men.[57] The influence of aging on postural response is further discussed in Chapter 4.

Grassi et al. reported that a decrease in cardiopulmonary pressure in healthy young human subjects results in increases in plasma NE concentrations averaging over 90% within 2 to 5 min; the plasma NE and forearm resistance changes had similar time and magnitude courses; the investigators noted that this cardiopulmonary receptor effect was in sharp contrast to the previously observed inability of the carotid baroreflex to alter these humoral and hemodynamic variables.[132] The above findings, taken together, suggest that in the human, cardiopulmonary reflexes play an important role in the regulation of vascular resistance during orthostasis. The important cardiovascular changes occurring during progression from low- to high-level LBNP have been well described by Rowell.[8]

D. GENERAL AUTONOMIC MECHANISMS IN ORTHOSTASIS

Orthostatic mechanisms have been analyzed with autonomic blocking agents.[133-136] Propranolol in normal adults did not materially alter the hemodynamic response to 60° head-up tilt.[134] Ferguson et al. found that after propranolol, baseline and low-level LBNP changes in forearm vascular resistance were relatively small but at –40 mmHg LBNP, there were reduced forearm vascular resistance responses, which were believed to be due to a propranol-induced decrease in the activity of the cardiac ventricular baroceptors.[135] After combined β-adrenergic parasympathetic blockade (propranolol-atropine), there was the expected lessening of the tachycardic response to 70° head-up tilt, with a greater than normal fall in cardiac output but no change in the mean arterial pressure response.[136] These studies indicate that in the normal human subject, the cardiac responses do not play a major role in the adaptation to the upright posture and strongly suggest that the peripheral vasoconstrictors are the critical elements in determining human orthostatic tolerance.

It has been noted by previous authors that heart rate changes from the supine to sitting position were usually less than from the sitting to standing position.[4,72,128] This was rather unexpected since the greater change in hydrostatic pressure at the carotid sinuses would be presumed to occur in the supine-to-sitting transition. Burke et al. investigated this question with the microneurographic technique, i.e., recording of the skeletal muscle sympathetic nerve activity (SMNA) in the limbs of awake subjects. He noted that pulse-synchronous bursts of peroneal muscle-nerve activity were triggered by arterial baroceptors and that the change in sympathetic burst activity per 100 heart beats was greater during supine to sitting than during sitting to standing.[72] However, the total multiunit sympathetic activity (MSA), i.e., the product of bursts per heart beat and heart rate, were about the same in supine to sitting as in sitting to standing. Further studies by Lindblad et al. indicated an inverse relation between MSA and heart rate levels in the supine position, but upon assumption of the upright position there was a direct relation between burst incidence per heart beat in the supine and the cardio-acceleration in the upright; they concluded that individual differences in postural heart rate change are related to individual MSA differences in the supine.[137] This significant relation between resting hemodynamic values and the corresponding values upon postural change is discussed further in Section IX below.

In an interesting recent report, Iwase et al. found that the tibial nerve-MSA minute burst rate had a close linear relationship to the sine of the tilt angle in graded head-up tilt in normal subjects.[71] Since it has been previously shown that a change in thoracic blood volume (TBV) in head-up tilt is closely related to the sine of the tilt angle[57] and that central venous pressure (CVP) is linearly related to TBV,[39] it may be that, aside from arterial baroceptors, cardiopulmonary baroceptors may also have an influence on skeletal muscle MSA. It is also possible that incremental gravity stimulation of somatosensory muscle and joint tissues during graded head-up tilt may also be involved in the change in MSA activity. It appears, however, that at present MSA cannot be equated with skeletal muscle vasoconstrictor activity and, as previously mentioned, some evidence suggests that these two are dissociated.[73]

Iwase et al., also found, as had been previously reported,[57,59] that with graded head-up tilt, heart rate increases exponentially, not linearly, i.e., the cardioacceleration is greater at the higher tilt angles.[71] Although heart rate and other cardiovascular responses are notoriously variable in LBNP,[138] carefully graded LBNP shows a similar rate change, i.e., a disproportionate cardioacceleration at higher stress levels.[130,131] This nonlinear heart rate response to gravity might suggest that heart rate changes during orthostasis are influenced not only by baroceptors, but also by other factors, e.g., the initial heart rate, by higher centers, or perhaps, as Iwase et al. have suggested,[71] by the exercise heart rate response,[79,80] which may depend on the different degrees of physical movement involved in the different postural changes.

E. SKELETAL MUSCLE PUMP (SMP)

Earlier investigators have stressed the importance of the SMP of the lower extremities in mitigating peripheral blood sequestration and in assisting venous return. Ludbrook showed that during quiet standing, three times more blood accumulated in the thigh than in the calf and at a much faster rate, due, he believed, to the greater capacity of intermuscular venous channels and less competent venous valves in the thigh.[14] Without directional valves, the SMP could not function. Prior studies have shown that valves are present at 2- to 4-cm intervals in the deep and superficial veins of the leg (even in veins of 0.08 to 0.15 mm in diameter) and are more numerous in deep than superficial veins. Valves become scarcer toward the heart and only 24% of external iliac veins have valves, and there usually are none in the common iliac vein or abdominal vena cava.[140] This would suggest that upon standing, the immediate reflux of venous blood that apparently occurs in the thigh, probably also occurs in the pelvis and buttocks.

Ludbrook also found that during quiet standing, the intramuscular pressures in different muscles of the lower extremities varied widely, but with a single maximal voluntary contraction, these pressures increased five- to ten-fold, i.e., to levels of 100 to 250 mmHg, varying with the individual muscle; such pressures approximated that of arterial pressure and confirmed previous reports that during maximal muscle contraction, blood flow to the calf is virtually abolished.[14] Other phases of Ludbrook's study indicated (1) that most of the calf blood sequestered during quiet standing is in intramuscular veins but most thigh blood was in intermuscular veins and (2) that a single maximal voluntary contraction of the combined thigh and calf muscles during standing could eject into the central circulation about 30% of the total blood volume that had previously accumulated upon assumption of the upright posture from recumbency.[14]

Anatomical studies have indicated that the ratio of venous to arterial blood volume is much greater in the extremities than in the internal organs. Superficial veins of the legs (greater or lesser saphenous) contain 95% of the total blood in skin and subcutaneous tissues, but deep leg veins (posterior and anterior tibial and peroneal) contain about 80% of the blood from underlying muscular structures. Each of the deep leg veins has a diameter about 35% greater than the adjacent artery, producing a combined deep venous volume that is about three times that of the corresponding arteries.[140,141] During vigorous bicycle exercise, it has been found that coincident with an ejection of 95 ml of blood centralward, there was a fall of about 50 mmHg of venous pressure in the leg veins; from these data, Ludbrook calculated that this venous pump was about twice as effective in conserving CBV as would be a maximum venoconstriction.[14] This provides perhaps a teleological answer to the previously posed puzzling question of why lower extremity venoconstrictor response is so paltry in the upright position.

Others have also investigated the effectiveness of voluntary lower limb muscle contractions. Rattan et al. found that electrical stimulation and volitional contraction of the lower limbs both resulted in 15 to 20% increases in stroke volume and cardiac output but the latter, unlike electrical stimulation, also increased heart rate, systolic blood pressure, and Vo_2.[142] While the heart rate and arterial pressure changes may be due to increased venous return, the volitional muscle contractions may also have been associated with the "central command" phenomenon, i.e., a cortical irradiation affecting central cardiovascular centers; this was suggested by Mitchell and Wildenthal[80] as responsible for the initial circulatory change in static exercise and may be the cause of the sudden arterial pressure increase upon standing.[79] Smith et al. found that moderate volitional lower limb tension during −50 mmHg LBNP increased heart rate and systolic blood pressure and eliminated presyncopal symptoms.[143] Asmussen demonstrated that

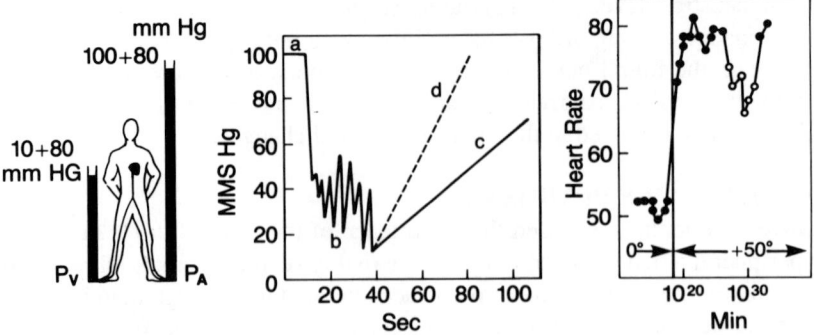

FIGURE 7. Effect of skeletal muscle pump action on lower extremity venous pressure and heart rate. (Left) Venous (P_V) and arterial (P_A) pressures in the foot in the erect posture. (From Rowell, L. D., *Human Circulation During Physical Stress,* Oxford University Press, London, 1986. With permission); (Middle) Effect of walking on ankle venous pressure (P_V): With the first step P_V drops from the quiet standing level of about 100 mmHg (a) to about 45 mmHg and fluctuates during walking (b). Upon resumption of quiet standing, pressure rapidly returns to the former level. The rate of return is shallow if the leg is cool (25°C), (c) but at a higher leg temperature (33°C) the rate of rise is doubled (d). (From Pollack, A. A. and Wood, E. H., *J .Appl. Physiol.* 1, 649, 1949. With permission.) (Right) Heart rate in supine position (0°) and during relaxation (closed circles) and voluntary contraction (open circles) during 50° head-up tilt. (From Asmussen, E., Christensen, E. W., and Nielsen, M., *Scand. Arch. Physiol.,* 81, 190, 1939. With permission.)

heart rate can be lowered by about 10 bpm if a subject in the head-up tilt position voluntarily contracts his or her leg muscles, as illustrated in Figure 7.[144] This may be a reflex bradycardia from a heightened arterial pressure. These findings raise the possibility that the saw-toothed arterial pressure and heart rate records common in prolonged standing or head-up tilt may be the result of periodic lower limb SMP activity.

In his review of the SMP, Laughlin reported that during the muscle contraction cycle, intramuscular pressures increase to 200 to 300 mmHg and the resultant compression of the veins propelled blood toward the heart; the subsequent relaxation permitted arterial influx. Capillaries and precapillary resistance vessels do not appear to be compressed during this process. The muscle pump adds kinetic energy to the system, and it has been estimated that about 30% of the energy necessary for the perfusion of skeletal muscles is provided by this pump during lower limb exercise.[145] As pointed out by Laughlin, because of the insertion of an energy-producing pump, the usual tenets of Poiseuille's law cannot be applied to this circuit without knowing the pump characteristics. The SMP may be more effective during locomotion than in vigorous exercise, because in locomotion the rhythmic recruitment of the muscle fibers is continuous and occurs in a wavelike fashion; this spatial and temporal sequence of pressure increase and decrease, as well as alternate lengthening and shortening of the muscle fiber during the push off and swing of the stride, is undoubtedly advantageous to the venous milking process.[145]

The studies of Pollack and Wood clearly showed that within the first few steps of walking, there is a rapid fall in ankle venous pressure from 100 mmHg to about 40 mmHg; the deep veins are compressed, blood is forced centrally, and reflux into the leg is prevented by the venous valves (Figure 7).[12] The return of venous pressure to relaxed levels usually takes about 30 s and the rate of return, which is a measure of the efficiency of the pump, is importantly influenced by the rate of arterial inflow. This efficiency of the SMP, as estimated by ankle venous pressure, appears to be highest at a treadmill rate of about 1.7 mi/h (slow walk).[12]

As stated above, a well-functioning venous valve system is an absolutely indispensable part of the SMP. Rowell pointed out that patients with absent or inadequate venous valves are prone to syncope and are very limited in their ability to increase stroke volume and cardiac output; temperature may also be a factor in SMP efficiency, since with combined exercise and heat stress, a shift of blood into dependent cutaneous veins may seriously reduce central venous filling pressures.[8]

Most of the previous discussion has dealt with the SMP during exercise or walking. A related and equally important question is the effectiveness of the SMP in the upright position during quiet standing. Previous investigators have marveled at the body's ability to withstand long periods of quiet standing.[13] There is considerable indirect evidence that postural sway (the unconscious slight body and leg movement during standing) is an indispensable element for maintenance of the vertical stance in humans. It is autonomously controlled by the short-circuit myotatic reflexes of the antigravity muscles.[13] Postural sway is an important factor in maintaining neuromuscular stability in the vertical position, which, as previously mentioned, is a difficult and mechanically disadvantageous one for the human, but its circulatory effect is probably of equal importance.

For the lower extremity muscles to serve as an effective pump during standing, it is necessary for the feet to anchor the body to the ground so that the muscles may have a counterweight against which to contract. If the body is suspended in an upright position with feet off the ground, this unconscious milking action of the lower extremity muscles is largely prevented. A 20- to 40-min suspension in this fashion will produce fainting in almost all subjects and a longer exposure is life threatening.[10] This great decrease of venous return is the rationale of death by crucifixion in which collapse and syncope were often hastened by the giving of wine, which caused vasodilation and further peripheral sequestration of blood.

There is other evidence of the importance of the SMP. Standing on a thick rug, e.g., creates minor instability of the lower extremities; this facilitates muscle pump action and makes standing more comfortable. Conversely, on very hard surfaces such as cement, stimulation of the plantar surfaces of the feet, from which the SMP reflex originates, is inhibited and SMP pump activity is much reduced. Prolonged standing on such surfaces is fatiguing because of the continued tendency towards a decreased cardiac output.

From studies of the muscular mechanics of standing, Nashner has found that functionally related postural muscles in the lower extremities are activated according to fixed patterns; muscular adjustments occur as a chain reaction, inevitably beginning at the fixed member (the feet in the case of stance) and progressing upwards to the knees and hips, so that leg and trunk muscles act in sequence in a preprogrammed manner.[146] It seems likely that this patterned neuromuscular sequence also activates the venous pump action of the limb and that the deprivation of the foot contact with the ground and the resultant lack of initiation of the ascending sequence of lower limb muscle righting reflexes is the cause of the circulatory collapse in bodily suspension above the ground.

As indicated in the foregoing and in the opinion of many reviewers, there seems little question but that the SMP of the lower extremities is a highly important factor in counteracting the gravity factor during both locomotion and quiet standing.[4,5,7,8,13,14,144,145] Rowell has called the SMP the "second heart"[8] and it seems likely that its actual importance in orthostasis has been seriously underestimated. A primary problem in studying this topic is the difficulty in quantitating the effect of the SMP. It would appear that increased effort to improve the methodology for the combined study of the SMP and venous valvular function in the upright posture would be of considerable value, not only for an improved understanding of postural physiology, but also for possible clinical application in patients with postural intolerance.

VII. EFFECTS OF LOWER BODY NEGATIVE PRESSURE (LBNP)

A. GENERAL

In the last 20 years a large research literature on LBNP has accumulated, particularly with regard to its use as an assay for orthostatic tolerance. LBNP consists of the application of negative pressure, usually in increments of about 10 mmHg, to a level of about −40 or −50 mmHg to the lower part of the body. This is accomplished by means of a vacuum pump attached to an air-tight box, which encloses the lower body below the iliac crests. As Wolthuis has pointed out in his extensive review, LBNP is a comfortable, controllable, reproducible, noninvasive method of bringing about a central hypovolemia. It has an advantage over postural stress, since it can be easily graded and rapidly released without the skeletal muscle activity incident to the assumption of the free-standing position, thus, minimizing motion artifacts. Because it can produce a central hypovolemia in a gravity-independent manner, this redistribution of blood volume can be studied during space flight, an added advantage.[147] It has proven valuable in the study of the effects of age, deconditioning, physical training, and other factors on this type of stress. LBNP will be further discussed in Chapters 2 to 5 of this review volume.

Since our primary interest in this chapter is in the physiological and pathological effects of the upright posture, we will not attempt to fully analyze the vast LBNP literature. However, since LBNP simulates postural stress in several, though not all, respects and produces somewhat similar physiological changes, we will review those aspects that are relevant to the study of circulatory effects of postural change and attempt to point out some differences in reactions to these two stresses.

LBNP to a level of about –40 to –50 mmHg will produce blood shifts very similar to those of the upright posture.[147] One of the main points of difference is that in LBNP the pressure change is transmitted to the superficial vessels first and is exerted in a tranverse, stepwise gradient from the exterior to the interior of the exposed parts of the body.[148] It has been estimated that about 95% of the pressure change is imposed on the subcutaneous tissue and about 80 to 90% on the underlying muscle,[149] so that the degree of pressure transmission is partially dependent upon the mass and characteristics of the underlying tissue. The LBNP pressure decrease is therefore transmitted to a variable degree in a cross-sectional direction to the vessels of the lower body from the pelvis to the foot, while in the upright posture there is an immediate, uniform, longitudinal pressure gradient in all vessels. Certain gravity effects do not occur in LBNP, e.g., the pulmonary blood flow redistribution, the SMP response, and the lesser transmural pressures at the baroceptor sites above the heart. As mentioned in Section IX, there is also evidence that at comparable degrees of stress (e.g., at –40 or –50 mmHg LBNP), the footward migration of the diaphragm, and perhaps also of the abdominal and thoracic organs, is less in LBNP than in the head-up posture[151].

B. REDISTRIBUTION OF BLOOD VOLUME
1. Upper Body Effects

Ebert et al. found a linear correlation between the decreases in thoracic blood volume (TBV) and degrees of LBNP in healthy young male subjects,[131,150] a finding similar to that reported during graded postural stress.[57] Wolthuis and his colleagues traced the fate of this redistributed blood volume by monitoring isotope activity over selected anatomical areas before, during, and after LBNP of –40 mmHg following injection of radioactive iodinated serum albumin.[45] This provided an approximation of the relative vascular volumes in the target areas. All upper body parts, except the splanchnic region, showed isotope depletion during LBNP. There was no change in density over the splanchnic region, thus confirming the previous findings of Sjostrand that there is little blood accumulation in the abdomen during orthostasis.[41] In Wolthuis' study there were, after injection of radioiodinated serum albumin (RISA), notable increases in isotope activity in all areas of the lower body; there were increases of 21% in the pelvis, 32% in the buttock, 35% in the thigh, 26% in the calf, and 7% in the heel. The percent pooling in the buttock and thigh were higher than in the calf; since the combined mass of the former two must be greater than that of the latter, the pooled blood volume percentage in buttocks and thighs is apparently much greater than previously suspected. It is also significant that the isotope activity curves in the thigh, buttock, and pelvis reach their peak in about $2\frac{1}{2}$ min, but the activity curves in the calf and the heel require about 5 min to reach their maximum.[45]

In the head-up posture there is an increase in pulmonary ventilation, a decrease in pulmonary blood volume, a redistribution of pulmonary blood flow, and changes in the ventilation to perfusion ratio;[16] during LBNP there are also pulmonary effects. At –40 mmHg LBNP, there is a diaphragmatic descent of about 1.7 cm (compared with a 3.5 cm descent in the head-up posture).[151] The decrease in central blood volume (CBV) and the downward movement of the diaphragm are undoubtedly responsible for the increase in total lung capacity (17%), functional residual capacity (35%), and residual volume (39%).[147] There is also a decrease in pulmonary capillary blood volume (32%) and a decrease in pulmonary diffusion capacity (22%) at –40 mmHg.[152] The diaphragmatic movement also results in a vertical shift of the cardiac electrical axis.

2. Dependent Pooling

The amount of dependent pooling during LBNP depends on the degree of negative pressure and the time of exposure. There was a reported pooling of approximately 0.5 to 0.6 l in the lower extremities at –40 mmHg[149] but at –70 mmHg sequestration increased to about 0.75 l.[148] Musgrave et al. conducted experiments in seven healthy young males subjecting them to levels of –20 to –40 mmHg of LBNP.[149] The volume of both legs (to the upper thigh) was monitored with a water plethysmograph and results were expressed as both relative and absolute volumes. The pooled volumes increased progressively, reaching about 80% of their maximum in $1^1/_2$ to 2 min; at 5 min a total volume of 614 ml had been pooled in the two extremities. These data (the results of repeated trials in the same subjects) were similar to those obtained in previous free-standing subjects.[149] Using the same technique, these investigators also tested two young male subjects with LBNP before and after 2 weeks of full bedrest deconditioning; after deconditioning the average leg volume after 5 min at –40 mmHg LBNP was 650 ml, compared with 590 ml before deconditioning; the difference was not significant.[153] There were also no changes in the venous distensibility curves of the lower extremities. However, the heart rate increase during LBNP was 60% greater after the study, suggesting that a normal deconditioning response had been elicited.[153]

The authors stated that the lack of change in lower extremity pooling after bedrest deconditioning was surprising, since previous head-up tilt experiments with astronauts had shown 87 to 89% leg-volume increase post-Gemini compared with pre-Gemini. On the other hand, preliminary data from Apollo 7 and 8 had indicated no significant pre- and postflight differences in blood volume pooled in response to LBNP. Since bedrest was assumed to induce physiological changes qualitatively similar to those of weightlessness, it was expected that there would be a similar response to LBNP. Among possible explanations, the authors suggested that there might be basic differences in head-up tilt and LBNP as an assay method.[153] In other LBNP studies on healthy young male subjects, Smith et al.[154] and Raven et al.[155] found after 5 min of exposure to graded LBNP at –50 mmHg, leg volume increases ranged from 300 to 310 ml (3.10 to 3.24%). Convertino et al. used a rheographic method to measure volume changes from midcalf to mid thigh in normal young male subjects at –50 mmHg LBNP; they noted an increase of 1.5 ml/100 ml of tissue or about a 2.8% increase in volume during exposure to LBNP.[156]

As Wolthuis et al. have pointed out, there is an obvious need for a technique that can provide a more quantitative measure of total dependent pooling during orthostatic stress; they also believed it to be highly desirable that successive experiments be repeated in the same subject.[147]

C. COMPENSATORY CIRCULATORY REACTIONS TO LBNP
1. Central Changes

A number of studies have been made on the central cardiovascular changes during LBNP.[8,131,147,150,154,160,163] After an exposure of 5 to 10 min at –40 to –50 mmHg , most investigators found heart rate increases ranging from 20 to 45%, [8,131,147,150,154,165] but others reported lesser increases of 6 to 10%.[160,163] Decreases in stroke volume and cardiac output usually ranged from 25 to 49% and 30 to 44%, respectively; forearm vascular resistance and total peripheral resistance usually increased from about 40 to 58% and from 25 to 40%, respectively. Systolic pressure was variable but usually decreased from 2 to 15%. Mean arterial pressure generally decreased from 2 to 12% and pulse pressure decreased from 10 to 25%. Diastolic pressure changes were variable, ranging from increases of 3% to decreases of 24%.[131,150,159,160,162] CVPs decreased by about 4 to 7 mmHg at –40 mmHg LBNP.[8,159,160,165]

There was usually a slight decrease in ventricular ejection fraction.[161] Ahmad et al. studied the reactions of 12 healthy men to –40 mmHg for 10 min; they found average decreases of about 22% in stroke volume, 20% in cardiac output, and 19% in end-diastolic volume, but no significant change in heart rate, blood pressure, ejection fraction, myocardial contractility or end-systolic volume.[163]

Tomaselli et al. reported cardiovascular hysteresis during LBNP; in many cases, the circulatory values during the descending phase were significantly different than those in the ascending phase, probably due to fluid sequestration in the interstitial and lymphatic compartments.[164] Nixon et al. found that the circulatory changes in LBNP were slightly greater than during head-up tilt, but the alterations in left ventricular end-diastolic volume and stroke volume were similar to those reported by others in LBNP and head-up tilt.[161] Nutter et al., after careful study of severe LBNP stresses in dogs, found no evidence of diminished myocardial contractility.[162] Wolthuis et al., in their extensive review of LBNP responses, emphasize that there were large variations in LBNP circulatory reactions; a good deal of the variation, they thought, was due to the diversity in the subjects used, differences in subject familiarization with the procedure and, most notably, the presence or absence of anxiety or emotional factors.[147] Variability in orthostatic responses is discussed further in Section IX below.

2. Peripheral Changes

In 1973 Montgomery et al. used impedance cardiographic methods to measure regional blood flows in different parts of the body. After 5 min of –40 mmHg, they recorded decreases of 34% in leg blood flow and 39% in pelvic flow.[157] Ebert et al. found at –10 mmHg LBNP, a small but definite increase in forearm vascular resistance, which leveled off at higher LBNP stresses. In young men, total peripheral resistance was increased at all levels of LBNP, but less so than in middle-aged and older men.[131] During low-level LBNP, muscle vessels achieved near-maximum vasoconstriction, but at increased LBNP, skin vessels showed a continuously progressive and graded vasoconstriction.[166]

Essendoh et al. recorded peripheral blood flow in both the forearm and calf during LBNP. At low levels (–20 mmHg), forearm blood flow was restricted more than calf flow, but at higher levels (–40 mmHg) or after a Valsalva maneuver, both forearm and calf vessels constricted markedly and to the same degree. The investigators concluded that reflex reduction of blood flow to the skeletal muscle of limbs resulting from the deactivation of low-pressure baroceptors is directed mainly to the arm.[68]

In a further effort to explain the unequal response of the forearm and calf blood flow to LBNP, two groups tested the efferent muscle sympathetic nerve activity (MSNA) in the limbs and found that MSNA was increased to a similar degree in the peroneal nerve as in nerves of the upper extremity.[167,168] A recent report by Anderson et al. showed that at low level LBNP (–20 mmHg) there was a decrease in brachial artery diameter as well as brachial artery flow with a small (3 mm) decrease in mean arterial pressure. During distal circulatory arrest (DCA) to an upper extremity, there was again a decreased brachial diameter (measured with ECHO) as well as a decreased forearm blood flow; with the addition of LBNP there was, however, no further decrease in flow to the extremity.[169] The authors concluded that LBNP did not cause direct reflex vasoconstriction of the brachial artery. Since the brachial artery flow during DCA was only about one-tenth that of the control, the results do not indicate that reflex constriction of brachial arterial vessels did not occur at a higher, more normal arterial flow level. However, the study does reemphasize the important hemodynamic principle that vascular resistance change, as calculated with Poisseuille's formula, is really only an abstraction and not necessarily a valid index of active vasomotor change if flow has been altered, as it usually is in orthostasis.[169] There is, however, little question that active vasoconstriction of muscle beds does occur during orthostasis, as indicated by the report of Ardill et al. that the decreases in forearm blood flow during LBNP were much less in a sympathectomized arm than in a control arm.[170]

Splanchnic blood flow decreases linearly, though moderately, at low-level LBNP; at higher levels (–50 mmHg) splanchnic blood flow decreased by 32% and splanchnic vascular resistance increased 30%.[8] At –60 mmHg LBNP there was a decrease of about 23% in effective renal plasma flow and a 19% decrease in the glomerular filtration rate, so that the decrease in the glomerular filtration rate is secondary to the decrease in renal plasma flow.[147] Hirsch et al. found that during prolonged, progressive LBNP (–10 to –40 mmHg) there was an initial decline and then a return of forearm blood flow and increases in the glomerular filtration rate and plasma renin activity; they confirmed previous reports of progressive decreases in central venous pressure (CVP) and splanchnic blood flow, but found decreases in renal blood flow only when arterial baroceptors (not cardiopulmonary) were unloaded.[171]

The mechanism of the circulatory changes in LBNP has been previously discussed, along with that of head-up posture, in Section VI above. Rowell stated that LBNP results in a widespread vasoconstriction, of which about 39% might be attributed to skeletal muscle and skin, about 33% to the splanchnic vascular bed, and the remaining 28% probably to the kidney; he also stated that the skeletal muscle, splanchnic, renal, and cutaneous responses to LBNP are similar to those in the head-up posture and that they result from both cardiopulmonary and arterial baroceptor reflexes. The cardiopulmonary reflexes are triggered by a 2- to 6-mmHg decrease in CVP and the arterial baroceptor reflexes by the arterial pressure changes, particularly by the 30% decrease in aortic pulse pressure.[8] A general comparison of the circulatory responses to LBNP and the upright posture is given in Section IX below.

VIII. PROLONGED ORTHOSTATIC STRESS AND FAINTING

A. NONINVASIVE MONITORING OF ORTHOSTATIC STRESS

Uncomplicated, postural stress (without invasive monitoring) for periods of 15 to 20 min generally evokes hemodynamic reactions very similar to the shorter term (5 min) effects described in Section V and shown in Figure 4, except that the 20-min head-up tilt or free stand in normal subjects is usually accompanied by a 5 to 10% incidence of presyncope or syncope. In the 1940s a number of investigators used free-stand and tilt-table tolerance (and other physical stresses) as determinants of fitness for military aviation and/or the ability to withstand G_z (head-to-seat) acceleration incident to aerobatics. Typical of these investigations was that of Graybiel et al.[172] In 1941 these investigators studied the response of 91 pilots, 18 to 49 years of age, to a 20- to 30-min 65° head-up tilt; 10% of these subjects fainted and another 14% had a "poor" reaction, i.e., early presyncope. The mean circulatory changes reported by these and other investigators employing similar protocols and using noninvasive methods were generally similar to those shown in Figure 4.

B. SEVERE ORTHOSTATIC STRESS

Beginning in the 1960s postural and LBNP studies were carried out to analyze the effect of physical conditioning, deconditioning (e.g., bedrest), and weightlessness on orthostatic tolerance. Others used these stresses to explore the characteristics and mechanisms of syncope. Many of these were invasive studies that heightened the fainting incidence. Using intraarterial and intracardiac catheters, Brigden et al. monitored healthy young subjects during 20 to 30 min of head-up tilt.[173] Forearm blood flow decreased sharply immediately upon tilting, but returned to about control levels in 5 or 10 min; several subjects experienced typical vasodepressor syncope with bradycardia, sweating, convulsions, and unconsciousness; recovery was prompt with the restoration of the supine position.[173] It is characteristic of orthostatic fainting that unless propped up by an ill-advised observer, full recovery is practically assured.[147]

Stevens and Lamb found in intravascularly instrumented subjects fainting percentages of 0, 58, 70, and 100% at LBNP levels of –25, –40, –60, and –80 mmHg, respectively.[160] Wolthuis divided syncopal episodes into two phases: (1) circulatory instability with phasic variations in blood pressure and heart rate but no change in forearm blood flow or venous tone and (2) a precipitous decrease in arterial pressure, heart rate, and forearm vascular resistance but little change in central venous pressure (CVP). He found that the fall in stroke volume, cardiac output, and arterial pressure usually occurred before the bradycardia.[147] Newberry et al. induced fainting in 29 normal subjects with an LBNP stress of –60 mmHg in the sitting position.[174] The heart rate rose and blood pressure fell, but both fell precipitously in the last 30 to 60 s before the faint. Forearm vascular resistance decreased in almost all cases; there was a 60% increase in pulmonary ventilation and about a 40% decrease in end-tidal P_{CO_2}. At higher ambient temperature (32°C) fainting occurred more rapidly and some subjects lost consciousness without developing bradycardia.[174]

Murray et al. studied several intravascularly instrumented, normal subjects who underwent graded hypovolemia to –60 mmHg LBNP until fainting.[165] Presyncopal symptoms such as sweating, yawning, sighing, restlessness, anxiety, and nausea often came on rather suddenly. Heart rate rose progressively but never exceeded 100 bpm. Mean arterial pressure was steady until shortly before syncope, but pulse pressure fell gradually (due to a decrease in systolic and increase in diastolic pressure) until collapse, at which point all the arterial pressures fell abruptly.[165] CVP (which was consistently the first reflection of hypovolemia) fell progressively to about 7 or 8 mmHg below baseline until about 20 min before syncope, at which point it either rose or leveled off. During the last 10 to 15 min, cardiac output, stroke volume, and central blood volume (CBV) fell progressively until LBNP was stopped.[165]

While studying 20 healthy males undergoing 70° head-up tilt, Ebert et al. found during the first 3 min the expected large, immediate decrease in thoracic blood volume (TBV) incident to the rapid peripheral sequestration of thoracic blood; this was followed by a gradual decline, due mainly to capillary diffusion.[39] However, two men who developed syncopal symptoms had abrupt further decreases in TBV (increases in transthoracic Z_0) at the time of fainting, amounting to about 70% of the initial rapid decline[39] (Figure 3A). This rather startling decrease in TBV may be due to a sudden fall in systemic venous pressure, resulting either from the collapse of arterial pressure or abrupt withdrawal of skeletal muscle pump (SMP) activity. It would be interesting to determine if this sudden fall in TBV is characteristic of hypovolemic syncope or of syncope in general.

C. ORTHOSTATIC FAINTING — PREDISPOSING AND ASSOCIATED FACTORS
1. Intravascular Instrumentation

Several investigators have clearly demonstrated that invasive methods predispose to fainting during orthostasis.[162,165,173,174] Stevens studied 233 pilots and navigators who underwent head-up tilt tests; he confirmed the oft-repeated observation of the very wide variation in tilt tolerance and fainting susceptibility; during 20-min head-up tilt there was a 9% fainting incidence with noninvasive sensors, a 10 to 20% incidence if special maneuvers such as breath holding or hyperventilation were added, and a 52% incidence with intraarterial and intravenous instrumentation. He believed that psychic factors were critical in the production of the faint.[175]

2. Ambient Temperature

Newberry et al. noted that LBNP fainting was enhanced at a higher temperature and humidity.[174] At comfortable environments, there was a steady fall in total peripheral resistance, blood pressure, and heart rate, but at high temperatures, the peripheral resistance remained at about twice the resting level and even at syncope was above baseline. In one subject, forearm blood flow came to a complete standstill before the faint.[174] Other investigators also noted a lessened orthostatic tolerance at higher temperatures.[4,7,8,13,54] Turner et al. reported a 46% greater leg volume increase with 15 min of head-up tilt in summer than winter, even though the leg volume was measured at thermoneutral temperatures.[54] Lind et al. found a much higher fainting rate with tilt at warmer temperatures and thought that, in this circumstance, fainting resulted from trunk pooling rather than dependent pooling; they also reported that at high temperatures, limb blood flows, except for skin, fell to very low levels.[176]

3. Hypocapnia

Early investigators reported hyperventilation in the upright posture and Turner et al. noted that overbreathing is more common among subjects with poor orthostatic tolerance.[54] Hyperventilation may serve as an auxiliary venous pump mechanism;[13] it may be due to ischemia of the medullary centers[7,13] or perhaps arise from proprioceptive impulses from lower body antigravity muscles incident to postural sway.[13] Thus it is theoretically possible that postural sway may serve as a respiratory stimulant and thereby as an auxiliary venous pump, as well as playing an important role in maintaining postural stability and serving as a SMP. However, we are not aware of any solid evidence on the possible role of respiration in augmenting venous return in the upright position.

During head-up tilt in normal subjects, there was — associated with the usual hyperventilation — a progressive decline in P_ACO_2 with average decreases of 4.0 to 5.1 mmHg below control values.[55,56,134] Newberry et al. found in normal subjects at −60 mmHg LBNP, a 60% increase in respiratory minute volume and a 41% decrease in end-tidal P_{CO_2}.[174] Thus, hyperventilation and hypocapnia occur in both postural and LBNP stresses. The potential importance of hypocapnia in fainting lies in the well-known strong cerebral vasoconstrictor action of decreased alveolar P_{CO_2}.[7,10] Only 1 or 2 min of active hyperventilation can reduce cerebral blood flow by about 40%[177] and cause periods of apnea, slowing of the EEG, and reduced consciousness and tetany.[178] Although uncomplicated hyperventilation rarely causes fainting in normal subjects, a fainting tendency initiated by other causes, e.g., postural change, may be precipitated by unnoticed hyperventilation. In such instances, clinical improvement of the syncope will result with inhalation of 5% CO_2 and 95% O_2.[179] These findings strongly suggest that hypocapnia may be a predisposing element in orthostatic fainting.

4. Neurohormones

In Section V.E it was noted that plasma concentrations of some neurohormones, particularly AVP, were increased only moderately during LBNP or head-up posture, but markedly with the onset of presyncope or syncope.[90,104,106] Taylor and Noble subjected 14 subjects to venesection; in seven who fainted, antidiuretic hormone (ADH) was found in the blood; in the seven who did not faint, no plasma ADH was found.[180] Since injection of posterior pituitary extract will apparently mimic the symptoms of fainting, it seemed that high ADH concentrations might initiate presyncopal symptoms.[181] On the other hand, recent studies have indicated that high AVP levels can be attenuated with antiemetic agents. The activation of central emetic pathways in the dorsal ventricular formation can be responsible for a high concentration of AVP; this, in turn, may stimulate gastrointestinal motility and food aversion.[182] So, apparently it is the nausea that instigates the increase in AVP and not the reverse.

5. Psychological Factors

Many investigators have commented on the association of certain psychological factors with stress. Shepherd and Mancia noted that marked alteration of heart rate, arterial pressure, and regional blood flow may follow both emotional and mental stress.[7] Long-term psychogenic stresses indigenous to the socioeconomic structure of modern society are believed to play an important role in the high incidence of certain circulatory diseases in the developed nations of the western world.[183] In orthostatic stress, the not infrequent occurrence of presyncope upon anticipation of the stress leaves little doubt of the importance of anxiety and other psychogenic factors in the genesis of fainting.[171,174,175] In blood centers, a 5% rate of presyncope or syncope in subjects undergoing a standard venesection of about 500 ml is common; an analysis indicated that it is three times as likely for previous "fainters" to have a presyncopal reaction than for randomly selected donors.[184] In another comparative study of fainter-prone individuals by Ruetz et al., it was noted that compared with control donors, fainters had a higher resting pulse rate, higher pulse pressure, and lower diastolic pressure; fainters also had lower alveolar P_{CO_2} tensions before and during a 500-ml venesection.[56] Fainters also showed significantly different responses to the Minnesota Multiphasic Personality Inventory test in two scale categories, i.e., hypochondria and depression; further analysis of the data indicated that fainters had a heightened awareness and overconcern with bodily function and a proneness to feelings of inadequacy and depression.[56] While there undoubtedly exists a small but definite percentage of evident fainter-prone individuals (about 3 to 5%),[184] it is likely that fainterproneness is a relative rather than an absolute characteristic. Furthermore it may well be that in some individuals variations in the fainting tendency may be based on physiological as well as psychological factors.

D. THE MECHANISM OF FAINTING

Although several factors undoubtedly predispose to vasovagal fainting during hypovolemic stresses such as orthostasis, investigators instinctively search for a final common pathway of the circulatory collapse. Since atropine will reverse the bradycardia but not the hypovolemia or fainting tendency, it is assumed that vasodilation (especially in skeletal muscle) is the main immediate factor in the syncope.[7,56] While this may be the ultimate cause of the final collapse, the initiating factors are probably diverse. Chief among several suggested initiating factors are (1) a ventricular reflex, (2) cerebral ischemia and (3) psychogenic factors.

1. The Ventricular Reflex

A continuing decrease in venous return, arterial pressure, and end-systolic volume with a high heart rate suggests that as the end-systolic volume approaches zero, the endothelial receptors of the contracting ventricle may be stimulated by compression. Afferent vagal impulses from this contracting ventricle might invoke a Bezold-Jarisch-type reflex and result in a depressed sympathetic outflow with a decreased heart rate and falling vascular resistance.[185] Wallin and Sundloff, while monitoring peroneal muscle sympathetic nerve impulses, noted in two subjects a sudden cessation of sympathetic efferent impulses to skeletal muscles with the onset of syncope;[186] this suggested that the fall in peripheral resistance was due to a cutoff of sympathetic constrictor influence and not the activation of vasodilators. They thought these findings to be consistent with the theory that left ventricular receptors might trigger vasovagal syncope in humans.[186]

2. Cerebral Ischemia

Since the brain is the center of consciousness, it is generally assumed that the ultimate cause of fainting in a normoxic environment is cerebral ischemia, but how cerebral ischemia is brought about is uncertain. As previously mentioned, hyperventilation and hypocapnia occur during both head-up posture[55,56,134] and LBNP,[171] and the strong cerebral vasoconstrictor action of hypocapnia is well known.[7,10,54,174] Normal adults in the supine position have a cerebral blood flow of about 35 to 60 ml/100 g of brain tissue per min; clinical evidence of cerebral ischemia appears rather consistently at flow rates of about 30 ml/100 g/min or less.[177] Under ordinary conditions, brain vessels have an unusual autoregulatory ability, but if the basic vascular tone is heightened, e.g., by hypocapnia or hypertension, autoregulation is less efficient and signs of ischemia, perhaps of the vasomotor center, may occur at lower levels of perfusion pressure.[179]

3. Psychogenic Factors

As mentioned earlier, anxiety and apprehension have been frequently associated with the fainting tendency.[56,147,174,175,184] In animal experiments, two higher centers have been particularly identified as important in central circulatory control. The "defense reaction area" in the limbic lobe has sympathetic outflow to elements of skeletal muscle, skin, kidney, and splanchnic areas; activation of this reaction pattern results in a series of responses that resemble the human reaction to anxiety, fear, mental work, and the anticipation of muscular exercise.[187] The second response pattern, described by Löfving, involves impulses from the adjacent sympathoinhibitory center of the limbic cortex, which markedly inhibit cardiosympathetic activity, excite cardiovagal fibers, decrease the tone of resistance and capacitance vessels of skeletal muscle, and thus profoundly increase muscle blood flow.[188] This response pattern, which occurs readily in the atropinized, vagotomized preparation, results in a sharp fall in blood pressure and has many characteristics of human vasodepressor syncope. There exist, therefore, in higher mammals two cortical centers with separate pathways capable of carrying out two distinct types of response to stress. Engel believes that the central theme of vasodepressor syncope is anxiety or fear, fear that the fainter must deny because appropriate action is not possible. The result is a strong affect of helplessness and a resignation to the stress, in effect, an opting for oblivion.[189] This type of fainting most often occurs in individuals who are overconcerned with their bodies. It has, therefore, been suggested that in the presence of such predisposing factors, e.g., hypovolemia, hypocapnia or cerebral ischemia, the "emotional faint" factor may become evident in the more anxious or fearful subject.[189]

IX. VARIABILITY IN ORTHOSTATIC STRESS RESPONSES

A. COMPARATIVE RESPONSES TO DIFFERENT ORTHOSTATIC STRESSES
1. Free Standing and Head-Up Tilt

As discussed in Section V.D the immediate (first 30 s) responses to free stand and to rapid head-up tilt are similar, but the immediate free-stand heart-rate complex is somewhat greater than that of the 70° head-up tilt.[77,79] In orthostatic studies of the 1950s and 1960s, some investigators supported subjects on the tiltboard with a saddle, presumably to minimize the skeletal muscle pump (SMP) action and thus perhaps simulate more closely the weightless state. Murray et al. found that the hemodynamic responses to the tilt table were similar with footboard and saddle suspension.[190] In most studies in which head-up tilt and free-standing experiments were done in the same subjects, heart rates increased 19 to 38% in head-up tilt and 30 to 41% in free stand, with a slight tendency toward a greater heart rate response in the latter stress.[54,64,191,192] In both head-up tilt and free stand, diastolic pressures increased from 13 to 21%, systolic pressures were unchanged, and pulse pressures decreased from 18 to 32%,[54,56,64] i.e., the results were generally similar. We found no comparative data on either cardiac output or stroke volume. Turner et al. reported that at 60° head-up tilt, O_2 consumption was increased about 6%, but during free stand it was increased 19% — three times more.[54] Theoretically the main functional difference between these two procedures would appear to be the SMP, which should be fully operative in the free stand, but probably less so in head-up tilt. This is another area in which quantitative methods for the SMP activity would, if available, be very helpful.

2. Head-Up Posture and LBNP

There are a number of similarities in the physiological responses of healthy young adults to the head-up posture and LBNP (usually compared at –40 to –50 mmHg). The changes in TBV appear to be comparable.[57,131,150] Lower extremity pooling (to the upper thigh) was similar at LBNP (–40 mmHg) and free stand.[149] However, in order to match the heart rate changes at 70° head-up tilt, it was necessary to use about –50 mmHg LBNP. Hyperventilation and hypocapnia occur with both posture stress[55,56,134] and LBNP.[174] The "initial heart rate complex" that occurs during the first 30 s after free stand and rapid head-up tilt was described in Section V.D; we are not aware of any similar immediate recordings of hemodynamic changes during LBNP, so a comparison is not yet possible.

Some differences have also been noted between the reaction to the head-up posture and LBNP. The physical differences in the direction and nature of the pressure change were discussed in Section VII.A. The degree of dependent pooling is reportedly similar during the stresses. However, as discussed in Section VII.C, above, Menninger et al. found that dependent pooling after bedrest was greater in the head-up posture than after LBNP; we are not aware of any other direct comparisons of dependent pooling induced by postural stress and LBNP after such deconditioning stresses.

The redistribution of pulmonary blood flow and alteration in the ventilation to perfusion ratios due to gravity[16] have not been determined after LBNP. Radiographic studies have suggested that the downward displacement of the diaphragm (and presumably of thoracic and upper abdominal organs as well) is greater in the head-up position than during LBNP.[151]

There also appeared to be some differences in systolic and diastolic pressure responses. In head-up tilt there was almost invariably a distinct increase in diastolic pressure ranging from 10 to 20%; [52,53,57,58,62,64] however, in LBNP the diastolic change varied from +3% to −24%.[115,131,134,139,144,146] While there was, with both stresses, a pronounced decrease in pulse pressure, this was usually achieved in head-up tilt by a steady systolic and a rising diastolic pressure, but in LBNP usually by a falling systolic pressure and little change or a fall in diastolic pressure. Assuming an approximately equal degree of dependent pooling and central hypovolemia, it might be theorized that a greater increase in diastolic pressure in the head-up position (and the need for a greater LBNP stress to match the heart rate than the dependent pooling responses) might be due to the lesser hydrostatic pressure at the baroceptor sites in the head-up position. As previously noted, there were, with both stresses, wide variations in hemodynamic responses, which was not surprising given the variety of experimental conditions, subjects, and protocols. In such a situation, it was particularly unfortunate that, except for Musgrave et al.,[193] there were, to our knowledge, no systematic comparisons of the overall hemodynamic responses to these two stresses in the same subjects, thus making a comparative analysis very difficult.

B. FACTORS INFLUENCING ORTHOSTATIC RESPONSES
1. General

Both individual factors (such as physical fitness, age, and sex) and environmental factors (such as weightlessness and bedrest) affect orthostatic responses and will be discussed in separate chapters. Fatigue and systemic disease reduce orthostatic tolerance,[4] and a negative correlation has been reported between subject height and postural tolerance;[194] because of the unquestioned effect of gravity on hydrostatic pressures during standing, it is rather surprising that the effect of the factor of subject height on postural hemodynamics has not been further investigated. Aschoff analyzed the diurnal response to 70° head-up tilt and reported the greatest lability of arterial pressure and heart rate at 3 a.m.; this was not meal dependent.[195] Postprandial hypotension, which influences hemodynamics, occurs mainly in the elderly.[196] In 21 subjects ($\bar{x}$ = 73 years), mean systolic pressure decreased about 11 mmHg and heart rate increased about 7 bpm within 1 to 2 h after a meal; there was also a highly significant inverse correlation in the postprandial period between the basal and sitting systolic pressures.[196]

2. Initial Hemodynamic Values

It has long been known that the initial resting state of vascular smooth muscle or of a vascular bed significantly affects its responsiveness.[197] While testing the response to carotid occlusion and to a variety of drugs in dogs, Hatch et al. found that the initial arterial pressure was positively correlated with the end pressure and negatively correlated with the pressure change, thus confirming the "law of initial values".[198] Schvartz and Meyerstein reported a high correlation between the reclining and orthostatic heart rate in normal young subjects[199] and Burke et al.[72] found that the magnitude of the peroneal MSNA in healthy subjects was inversely related to the change in burst incidence upon assumption of the sitting and standing posture. In healthy young males, Smith et al. found significant, positive correlations between supine control systolic, diastolic, pulse, and mean arterial pressures and their values during 70° head-up tilt and free stand; there were also significant negative correlations between the supine values and the change in these values to 70° tilt and free standing.[64]

The mechanism of these intraindividual correlations is not known, but may be related to the classic baroceptor feedback phenomenon and/or to the baseline cardiac and vasomotor tone. It is known, for example, that the cardiac response may depend upon the existing balance of sympathetic and parasympathetic control at the time of stimulation, as shown by Robinson et al.[129]

C. VARIABILITY IN ORTHOSTATIC RESPONSE

All biological experiments are plagued with inter- and intrasubject response differences; this "physiological" variability makes it difficult to differentiate a true from a random response and often frustrates even the best experimental design. Responses to LBNP show great variances. Wolthuis et al. found large response variabilities to –40 mmHg LBNP in heart rate and blood pressure, both between and within subjects and in both baseline and delta values. They concluded that the LBNP stress level should be tailored to the individual.[138] Hyatt et al. studied heart rate responses to 70° head-up tilt, free stand, and –30 to –50 mmHg LBNP; some subjects showed consistent responses, others much variability.[192] After three tests, LBNP had the greatest day-by-day heart rate variability and they questioned its use as an assay technique; however, because the average heart rate changes in response to bedrest were greatest in LBNP, they concluded it was the most sensitive index of bedrest change.[192] Lower limb volume changes to LBNP showed, however, relatively little variability.[138,149]

Wolthuis et al. tested responses of 19 normal subjects to 5 min of quiet standing with ten retrials of each subject at 2-week intervals. Some subjects showed consistent responses and others did not. The within-subject estimate of variance (one standard deviation) of a single blood pressure reading in either position was about 6 mmHg.[200] Turner et al. reported considerable variability in the hemodynamic response to the head-up position in normal young women.[54] Schvartz studied test-retest reliability in 18 normal individuals during 20-min head-up tilts repeated 7 d apart;[201] he found the highest reproducibility in orthostatic systolic pressure readings (r = 0.85); the orthostatic-diastolic and pulse-pressure readings were slightly less reliable. The lowest test-retest reliabilities were found in the supine and the delta values of blood pressure.[201] In contrast to the very wide heart-rate variability in the stabilized supine and head-up posture, Ewing et al. found the group mean values in the immediate heart rate response (60 s) to the head-up posture were very reproducible on successive determinations done at short intervals.[77] With the microneurographic method, Burke et al. found that each individual had a fairly reproducible level of muscle sympathetic burst activity from day to day, but there were marked interindividual differences; in the sitting position, the wide intersubject differences were somewhat reduced, suggesting that sympathetic neurons are subjected to a fairly homogeneous central drive.[72]

It has been noted that subjects vary considerably in the means by which they maintain their arterial pressure during head-up tilt. In one study, some subjects showed little decrease in stroke volume and cardiac output and very little increase in vascular resistance; others had marked decreases in cardiac output and large increases in vascular resistance.[53] Other investigators recently tested the responses of healthy young males to a combined head-up tilt-LBNP stress and separated "vascular-type" subjects, who responded mainly through an increase in total vascular resistance, and "cardiac types", who responded mainly through an increase in heart rate. During a 20-d bedrest deconditioning experiment, the "vascular types" had many fewer symptoms than the "cardiac types", suggesting that the former adapted better to a hypokinetic environmental stress.[202]

It will be recognized that the accounts of variability in the studies quoted above are not as precise as would be desired, reflecting, in some cases, the absence of quantitative measures of test-retest reliability. The reports do, however, clearly indicate a wide variability in the hemodynamic response to orthostasis, which is certainly a formidable barrier to experimental progress in this field. It would seem evident that a greater emphasis on repetition of stresses using identical methodology along with inter- and intraindividual comparisons, as done in some studies,[72,200,202] would be very helpful, not only in delineating hemodynamic mechanisms, but perhaps in providing reasons for the wide variability in these orthostatic responses.

D. PREDICTORS OF ORTHOSTATIC TOLERANCE

There are different opinions on the optimal predictive indices of orthostatic tolerance, not only because of the different types and degrees of orthostatic stress that have been used, but also because of differing objectives; some investigators use orthostatic tolerance tests as indices of physiological fitness, some as a measure of the effect of deconditioning or weightlessness, and others as early indicators of pathological change. The signs that correlated best with a collapse tendency were a high recumbent[52,60,147,203] or high initial tilt heart rate,[203] a declining systolic or pulse pressure,[172,203] lesser muscle strength,[194] greater physical height,[195] and greater maximum leg volume during tilt.[203] Heart rate is apparently the most sensitive indicator of orthostatism associated with bedrest and the zero-gravity state,[204,205] which, in turn, are usually associated with a decrease in blood volume or CBV.[205] Cardiac output, peripheral resistance, and regional blood flow data, which would be most valuable as signs of impending orthostatic collapse, are unfortunately not available.

X. SUMMARY AND CONCLUSIONS

A. GENERAL

The circulatory handicap of the erect position is mainly twofold: (1) about 10% of the blood volume is immediately dislocated from the thorax to the lower body and thus temporarily unavailable to the heart for pumping purposes and (2) in the erect position the head lies well above the heart, which must now pump blood against the hydrostatic gradient to provide adequate blood flow to the brain.[4,5] To counteract these influences requires complex interaction of autonomic reflexes and a well-functioning circulation. Postural sway, which is characteristic of the human, provides an important assist in this process by activating the skeletal muscle pump (SMP). In this chapter, (1) the physical and physiological factors that influence the postural response, have been described, (2) the mechanisms involved have been analyzed, and (3) the factors that influence the variability of the response have been discussed. We have used the general term *orthostatic* stress to refer not only to the upright position itself, but also to lower body negative pressure, a technique often used to simulate the upright position.

As a result of the head-to-foot pressure gradient, vascular pressures in vessels of the lower part of the body rise, and those of the upper part fall. Between these two, at a neutral, transverse plane, about 10 to 15 cm below the heart, is the hydrostatic indifferent point, at which intravascular pressures do not change with posture.[4,5,9] This location is significant since it means that in the upright position all cardiac pressures are negative.

B. METHODS

For orthostatic studies, the most common invasive methods for cardiac output have been dye dilution and thermodilution,[20] and the most common noninvasive methods have been impedance cardiography[24-30] and CO_2 rebreathing.[21] Limb volume is usually measured by strain-gauge plethysmography.[40] Although thoracic (central) blood volume is difficult to measure, some determinations have been made by estimation of the volume between the injection site and the sampling site using dye dilution.[31-33] Transthoracic impedance (Z_0) provides valid estimates of changes in (but not absolute values of) thoracic blood volume (TBV) as shown by previous studies in the dog[23,34-36] and man.[37-38]

C. BLOOD VOLUME DISPLACEMENT DURING HEAD-UP POSTURE

In recumbent, resting humans, the total blood volume is about 70 to 75 ml/kg.[41,42] About 30 to 35% of this total, i.e., about 25 ml/kg, is thoracic (central) blood volume.[4,31] In the head-up posture, blood is dislocated to the lower body in two stages. The first or fast phase begins immediately and is about 90% complete in 2 to 3 min. In this phase about 7 to 10 ml/kg of blood migrates to the lower body within 5 min. About 90% of this blood comes from the thorax. The second or slow phase involves transcapillary diffusion of cell-free and protein-free fluid into the interstitial tissues. Rough estimates indicate that from the 5th to 20th min about 3 ml/kg of plasma filtrate is sequestered in the lower body. The diffusion rate, rapid at first, apparently decreases exponentially. There is some evidence that about 80% of the thoracic blood is translocated to the pelvis, buttocks, thighs, and legs,[14,41,147] and that while the pooled venous blood in the feet and legs is arterial in origin, that in the thigh and pelvis, because of the paucity of venous valves, may be primarily venous reflux blood.[14] The two phases of blood volume dislocation are shown graphically in Figure 3.

The rate and amount of blood dislocated from the thorax in the head-up position is the best single measure of the degree of orthostatic stress. This is illustrated by the fact that the degree of peripheral pooling,[13] stroke volume, cardiac output, blood pressure, vascular resistance changes,[57,59] and muscle sympathetic nerve activity (MSNA)[71] all have a predictable relationship to the sine of the tilt angle, which is itself directly related to the amount of dislocated thoracic blood.[57] As a consequence, we believe that an important goal for future investigation should be to increase efforts to quantitate the rate and degree of thoracic blood volume (TBV) dislocation, and particularly to develop noninvasive methods to achieve this purpose.

D. COMPENSATORY RESPONSES OF CIRCULATORY SYSTEM

In the healthy individual, the immediate compensatory responses to peripheral pooling are very effective; central arterial pressures are well maintained and blood flow to key organs and tissues, though reduced, is not seriously compromised. Graded head-up tilt studies show that in healthy young subjects TBV and ventricular stroke volume decrease linearly,[57,59] as shown in Figure 6. Calf volume also increases linearly with increasing gravity increments.[13] Cardiac output decreases progressively during lower levels of tilt, but then levels off at higher gravity levels.[57,59] Within 3 to 5 min after the assumption of the head-up position (or 70° head-up tilt), there are decreases in stroke volume (30 to 45%), cardiac output (18 to 22%), and pulse pressure (14 to 22%). During this period there are also increases in A-V oxygen differences (1 to 2 vol%), in heart rate (30 to 35%), systolic pressure (0 to 5%), diastolic pressure (12 to 15%), mean arterial pressure (2 to 10%), and total peripheral resistance (30 to 40%).[22,53,57,59,62-65] Typical central hemodynamic responses to the upright posture are shown graphically in Figure 4.

There is decreased flow to practically all peripheral organs and tissues, especially skeletal muscle (40 to 60%),[68] splanchnic area (40%),[8] and kidney (50 to 60%).[4-8] There is a decrease of about 15 to 20% in cerebral blood flow.[4,5,8,19] Venoconstriction is limited and transitory.[6] There is an "initial heart rate complex", which occurs upon standing or fast head-up tilt. It is characterized by a sharp 30 to 35% increase in heart rate, which peaks at 13 to 16 s, then falls rapidly.[74-79] Since this rise is much diminished in peripheral neuropathy, the heart rate complex is used as a diagnostic test for autonomic pathology[79,81] (Figure 5).

E. NEUROHORMONAL RESPONSES

Several neurohormonal systems are activated in the upright posture or during LBNP. There are rapid and significant increases in plasma catecholamines,[84,87-91] activation of the renin angiotensin system,[89,94-96] arginine vasopressin (AVP),[89] (especially during presyncope or syncope),[90,100,103-105] and decreased activation of atrial natriuretic factor (ANF).[109,110,116,117] In general, neurohormones seem to play only a minor role in the rapid hemodynamic adjustment to the upright posture. However, AVP, ANF, and other hormones are undoubtedly important in long-term fluid volume and other aspects of circulatory regulation.

F. MECHANISM OF CIRCULATORY RESPONSE TO HEAD-UP POSTURE

Several mechanisms are rapidly and simultaneously activated to bring about the compensatory responses. Two of the most important are (1) autonomic reflexes[4-8,121,128] and (2) the skeletal muscle pump (SMP) of the lower body.[4,14]

1. Autonomic Reflexes

Autonomic reflexes are undoubtedly the key factors determining rapid hemodynamic adjustment to the upright position. While it is assumed that the arterial rather than the cardiopulmonary baroceptors play a major role in the heart rate increase, the relative roles of the two systems are not easily determined, since central venous pressure (CVP) and arterial baroceptor pressures both decline during orthostasis. Both pharmacological and neck suction-neck pressure studies indicate that arterial baroreflexes are important determinants of acute orthostatic heart-rate changes. Cardiosympathetic and parasympathetic blockade have relatively little effect on the general response to the head-up posture,[133-136] suggesting that hemodynamic postural intolerance is primarily attributable to a deficiency in peripheral vascular resistance.

Lower degrees of graded head-up tilt[57] or LBNP[130,131] (which do not affect heart rate or arterial pressure) increase total vascular resistance to a level of 70% or more of that achieved at full orthostatic stress; these and other findings strongly suggest that in the human the cardiopulmonary baroceptors play an important role in regulating vascular resistance during orthostasis. There is evidence, however that arterial baroceptors may be important in splanchnic vasoconstriction.[8]

2. Skeletal Muscle Pump (SMP)

There is strong but mainly indirect evidence that in the upright posture the SMP is a vital factor in assisting venous return.[4,5,8,14] During walking and running, the importance of this mechanism in pumping blood centrally is unquestionable[8,12] (Figure 7). What is less clear and needs further study is the role of the SMP during the postural sway of quiet standing. The evidence indicates that the continuous sway movements, which involve trains of somatic proprioceptive reflexes beginning at the plantar surfaces of the feet, cause alternate waves of contraction and relaxation of the entire antigravity muscle system. These reflexes not only provide postural stability, but are essential in maintaining venous return in the upright posture.[8,13,14,145,146] We believe that the importance of the SMP in counteracting the orthostatic gravity effect has been considerably underestimated and that further investigation of the SMP might provide important leads for the diagnosis and treatment of postural hypotension and related orthostatic disorders.

G. LOWER BODY NEGATIVE PRESSURE (LBNP)

LBNP is a controllable, noninvasive method of producing central hypovolemia, which simulates, in many respects, the head-up posture.[147] They both induce, for example, decreases in thoracic blood volume (TBV) and central venous pressure (CVP),[57,131,150] blood pooling in the lower extremities,[149] decreases in stroke volume and cardiac output, increases in vascular resistance,[8,53,147,154,155,193] and hyperventilation and hypocapnia.[55,56,134,174] Primary differences include the direction and nature of the pressure change and tendencies toward greater increases in heart rate and diastolic pressure in the upright posture; the latter differences may be due to the lesser hydrostatic pressure at the baroceptor sites in the upright position. Other differences include the pulmonary blood flow changes, ventilation-perfusion redistribution, and SMP activity in the upright position.[153]

LBNP is widely used to simulate postural stress. Yet, in spite of the physiological differences between them, direct comparisons of these two stresses are almost nonexistent. If LBNP is used as an index of postural stress, we believe it should be preceded or accompanied by a systematic comparison between the two, in the same subjects and under comparable conditions. Such studies might contribute significantly to an understanding of the differences in mechanisms and perhaps provide useful leads to possible interventions for controlling their respective circulatory changes.

H. PROLONGED ORTHOSTATIC STRESS AND FAINTING

Uncomplicated postural stress of 15 to 20 min, noninvasively monitored, evokes hemodynamic responses similar to those described above (Figure 4). However, prolonged head-up stress of 20 to 40 min or extended LBNP at −50 to −70 mmHg often result in decreases in systolic and pulse pressure, large increases in heart rate, and an increasing incidence of vasovagal syncope. Intravascular instrumentation markedly accelerates this tendency.[147,160,165,174] Increased ambient temperature,[4,7,8,13,54,174,176] psychological factors such as anxiety,[172,174,175] and the hypocapnia accompanying hyperventilation[55,56,174] also increase the incidence of syncope.

It is generally assumed that sudden vasodilation, especially of vessels of skeletal muscles, is the final common pathway of the fainting mechanism.[56] It is theorized that the initiating mechanism may be either (1) a ventricular reflex orginating at the endothelial surface of a strongly contracting but relatively empty ventricle,[185] (2) cerebral ischemia,[56] or (3) psychogenic factors, acting through the sympathoinhibitory center of the limbic lobe.[56,188,189]

I. FACTORS INFLUENCING THE ORTHOSTATIC RESPONSE

Several studies have indicated that freestanding and 70° head-up tilt produce similar hemodynamic changes and approximately similar degrees of stress.[54,64,192] Physical fitness, age, sex, and weightlessness importantly influence orthostatic response and are discussed in separate chapters of this review. Fatigue,[4] systemic disease, and greater subject height[194] decrease orthostatic tolerance. A higher supine heart rate, or a higher heart rate or lower pulse or systolic pressure during early tilt, are predictive of low postural tolerance.[52,60,147,203-205]

Previous studies indicate that hemodynamic responses to the upright posture are subject to the "law of initial values",[197,198] i.e., a tendency toward a negative correlation between the reclining and delta orthostatic arterial pressures[64] or heart rates.[199] In a microneurographic study, the magnitude of the supine peroneal muscle sympathetic nerve activity in healthy subjects was also inversely related to the change in burst incidence upon assumption of the sitting and standing position.[72]

J. RESPONSE VARIABILITY

Prior studies have shown large inter- and intraindividual variability in responses to postural[54,200,201] and LBNP[138,192] stresses. There are also wide interindividual differences in supine muscle sympathetic nerve activity (MSNA) and changes in burst activity with postural change.[72] The reason for this diversity is unknown. In two studies of head-up tilt, it was noted that different subjects maintained their blood pressure in different ways, some primarily through adjustment of vascular resistance, and others by greater adaptation of their cardiac responses.[53,202] This suggests the possibility that further analysis of orthostatic hemodynamic manifestations might be helpful, not only in clarifying basic response mechanisms, but also in helping to explain the wide interindividual variability in orthostatic responses. It is also likely that more intersubject (in addition to group) analyses that have proven fruitful in certain studies[72,138,200] would also contribute substantially to our understanding of postural physiology.

REFERENCES

1. **Ember, C. R. and Ember, M.,** Early hominids and their cultures, in *Anthropology,* 5th ed., Prentice-Hall, Englewood Cliffs, NJ, 1988, 62.
2. **Campbell, G. G.,** The evolution of hominid behavior, in *Humankind Emerging*, Scott, Foresman, Glenview, IL, 1988, 225.
3. **Martin, J. P.,** *The Basic Ganglia and Posture*, Lippincott, Philadelphia, 1967.
4. **Gauer, O. N. and Thron, H. L.,** Postural change in the circulation, in *Handbook of Physiology*, Sect. 2, *Circulation*, Vol. 3, Hamilton W. F. and Dow, P. Eds., American Physiological Society, Washington, D. C., 1965, 2409.
5. **Blomqvist, C. G. and Stone, H. L.,** Cardiovascular adjustments to gravitational stress, in *Handbook of Physiology*, Sect. 2, *The Cardiovascular System*, Shepherd, J. T., Abboud, F. M., and Geiger, S. R., Eds., American Physiological Society, Bethesda, MD, 1982, 1025.
6. **Rothe, C. F.,** Reflex control of veins and vascular capacitance, *Physiol. Rev.,* 63, 1281, 1983.
7. **Shepherd, J. T. and Mancia, G.,** Reflex control of the human cardiovascular system, *Rev. Physiol. Biochem Pharm.,* 105, 1, 1986.
8. **Rowell, L. D.,** *Human Circulation During Physical Stress*, Oxford University Press, London, 1986.
9. **Kirsch, K. A. and von Ameln, H.,** Some functional characteristics of the low pressure system, in *Oxygen Transport to Human Tissues*, Leoppky, J. A. and Riedesel, M. L., Eds., Elsevier/North Holland, New York, 1982.
10. **Smith, J. J. and Kampine, J. P.,** *Circulatory Physiology*, 2nd ed., Williams & Wilkins, Baltimore, 12, 1984.
11. **Shepherd, J. T. and Van Houtte, P. M.,** *Veins and their Control*, W. B. Saunders, London, 1975.
12. **Pollack, A. A. and Wood, E. H.,** Venous pressure in the saphenous vein at the ankle in man during exercise and change in posture, *J. Appl. Physiol.,* 1, 649, 1949.
13. **Hellebrandt, F. A. and Franseen, E. B.,** Physiological study of vertical stance in man, *Physiol. Rev.*, 23, 220, 1943.
14. **Ludbrook, J.,** Musculovenous pump of human lower limb, *Am. Heart J.,* 71, 635, 1966.
15. **Avasthey, P. and Wood, E. H.,** Intrathoracic pressure relations during response to change in body position, *J. Appl. Physiol.*, 37, 166, 1974.
16. **West, J. B.,** *Respiratory Physiology*, 2nd ed., Williams & Wilkins, Baltimore, 1979.
17. **Strandgaard, S. and Paulson, O. B.,** Cerebral auto-regulation, *Stroke*, 15, 413, 1984.
18. **Busija, D. W. and Heistad, D. D.,** Factors involved in the physiological regulation of the cerebral circulation, *Rev. Physiol. Biochem. Pharm.*, 201, 161, 1984.
19. **Scheinberg, P. and Stead, E. A.,** Cerebral blood flow in male subjects, *J. Clin. Invest.*, 28, 1163, 1949.
20. **Rothe, C. F.,** Fluid dynamics, in *Physiology,* Selkurt, E. E., Eds.,Little, Brown, Boston, 1984.
21. **Farhi, L. E.,** Dilution methods for measurement of cardiac output: a review, in *Oxygen Transport to Human Tissues*, Leoppky, J. A. and Riedesel, M. L., Eds., Elsevier/North Holland, New York, 1982, 32.
22. **Loeppky, J. A., Greene, E. R., Hoekenga, D. E., et al.,** Beat by beat stroke volume assessment by pulsed Doppler in upright and supine exercise, *J. Appl. Physiol.*, 50, 1173, 1981.
23. **Geddes, L. A. and Baker, L. E.,** *Principles of Applied Biomedical Instrumentation*, 2nd ed., John Wiley & Sons, New York, 1975, 320.
24. **Kubicek, W. G., Kottke, F. J., Ramos, M. U., et al.,** The Minnesota impedance cardiograph — theory and applications, *Biomed. Eng.*, 9, 410, 1974.
25. **Denniston, J. C., Maher, J. T., Reeves, J. T., et al.,** Measurement of cardiac output by electrical impedance at rest and during exercise, *J. Appl. Physiol.*, 40, 91, 1976.
26. **Ebert, T. J., Eckberg, D. L., Vetrovec, G. M., et al.,** Impedance cardiograms accurately estimate changes in left ventricular stroke volume in man, *Cardiovasc. Res.*, 18, 354, 1984.
27. **Edmunds, A. T., Godfrey, S., and Tolley, M.,** Cardiac output measured by transthoracic impedance cardiography, *Clin. Sci.*, 63, 107, 1982.
28. **Muzi, M., Ebert, T. J., Tristani, F. E., Jeutter, D. C., Barney, J. A., and Smith, J. J.,** Determination of cardiac output using ensemble-averaged impedance cardiograms, *J. Appl. Physiol.,* 58, 200, 1985.
29. **Miller, J. C. and Horvath, S. M.,** Impedance cardiography, *Psychophysiology,* 15, 80, 1978.
30. **Miles, D. S. and Gotshall, R. W.,** Impedance cardiography: non-invasive assessment of human central hemodynamics at rest & during exercise, *Exerc. Sport Sci. Rev.*, 17, 231, 1989.
31. **Yu, F. N.,** *Pulmonary Blood Volume in Health and Disease*, Lea & Febiger, Philadelphia, 1969.
32. **Marshall, R. J. and Shepherd, J. T.,** Interpretation of changes in "central" blood volume and slope volume during exercise in man, *J. Clin. Invest.*, 40, 385, 1961.
33. **McIntosh, H. D., Gleason, W. L., and Miller, D. E.,** Major pitfalls in the interpretation of "central" blood volume, *Circ. Res.*, 9, 1223, 1961.
34. **Luepker, R. V., Michael, J. R., and Warbasse, J. R.,** Transthoracic electrical impedance: quantitative evaluation of a non-invasive measure of thoracic fluid volume, *Am. Heart J.*, 85, 83, 1973.

35. **Patterson, R. P., Kubicek, W. G., Witsoe, D. A., et al.,** Studies on the effect of controlled volume change on thoracic electrical impedance, *Med. Biol. Eng. Comput.*, 16, 531, 1978.
36. **Denniston, J. C. and Baker, L. E.** Measurement of pleural effusion by electrical impedance, *J. Appl. Phys.*, 38, 851, 1975.
37. **Ramos, M. U., LaBree, J. W., Remole, W., and Kubicek, W. G.,** Transthoracic electric impedance: a clinical guide for pulmonary fluid accumulation, *Minn. Med.*, 58, 671, 1975.
38. **Van de Water, J. M., Mount, B. E., Barela, J. R., et al.,** Monitoring the chest with impedance, *Chest*, 64, 597, 1973.
39. **Ebert, T. J., Smith, J. J., Barney, J. A., Merrill, D. C., and Smith, G. K.,** The use of thoracic impedance for determining thoracic blood volume in man, *Aviat. Space Environ. Med.*, 57, 49, 1986.
40. **Englund, N., Hallbreuk, T., and Ling, L. G.,** The validity of strain gauge plethysmography, *Scand. J. Clin. Lab. Invest.*, 29, 155, 1972.
41. **Sjostrand, T.,** Volume and distribution of blood and their significance in regulating the circulation, *Physiol. Rev.*, 33, 202, 1953.
42. **Levinson, G. E., Pacifico, A. B., and Frank, M. J.,** Studies of cardiopulmonary blood volume, *Circulation*, 33, 347, 1966.
43. **Milnor, W. R.,** Pulmonary Circulation, in *Medical Physiology*, Vol. 2, 14th ed., Mountcastle, V. B., Ed., C. V. Mosby, St. Louis, 1980.
44. **Asmussen, E.,** Distribution of blood between the lower extremities and rest of body, *Acta Physiol. Scand.*, 5, 31, 1943.
45. **Wolthuis, R. A., LeBlanc, A., Carpenter, W. A., et al.,** Response of local vascular volumes to LBNP stress, *Aviat. Space Environ. Med.*, 46, 697, 1975.
46. **Hagan, R. D., Diaz, F. J., and Horvath, S. M.,** Plasma volume changes with movement to upright position, *J. Appl. Physiol.*, 45, 414, 1978.
47. **Hinghofer-Szalkay, H. and Moser, M.,** Fluid and protein shifts after positional changes in humans, *Am. J. Physiol.*, 250, H68, 1986.
48. **Thompson, W. O., Thompson, P. K., and Dailey, M. E.,** Effect of posture on the composition of blood in man, *J. Clin. Invest.*, 5, 573, 1928.
49. **Tarazi, R. C., Melsker, H. J., and Dunstan, H. P.,** Plasma volume changes with upright tilt, *J. Appl. Physiol.*, 28, 121, 1970.
50. **Mellander, S., Oberg, B., and Delram, H. O.,** Vascular adjustments to increased transmural pressure in cats and man, *Acta Physiol. Scand.*, 61, 34, 1964.
51. **Kirsch, K. A., Merke, J., and Hinghofer-Szalkay, H.,** Fluid volume distribution during changes in body posture in man, *Pflügers Archiv.*, 383, 195, 1980.
52. **Asmussen, E., Christensen, E. H., and Nielsen, M.,** The regulatory circulation in different postures, *Surgery*, 8, 604, 1940.
53. **Smith, J. J., Bush, J. E., Wiedmeier, V. T., and Tristani, F. E.,** Application of impedance cardiography to the study of postural stress in the human, *J. Appl. Physiol.*, 29, 133, 1970.
54. **Turner, A. H., Newton, M. J., and Haynes, F. W.,** The circulatory reaction to gravity in healthy young women, *Am. J .Physiol.*, 94, 507, 1930.
55. **Anthonisen, N. R., Bartlett, D., and Tenney, S. M.,** Postural effects on ventilatory control, *J. Appl. Physiol.*, 20, 191, 1965.
56. **Ruetz, P. P., Johnson, S. A., Callahan, R., Meade, R. C., and Smith, J. J.,** Fainting: a review of its mechanisms and a study in blood donors, *Medicine*, 46, 363, 1967.
57. **Smith, J. J., Hughes, C. V., Ptacin, M. J., Barney, J. A., Tristani, F. E., and Ebert, T. J.,** Hemodynamic response to graded postural stress in normal males: the effect of age, *J. Gerontol.*, 42, 406, 1987.
58. **Stafford, R. W., Harris, W. S., and Weissler, A. M.,** Left ventricular systolic time intervals as indices of postural circulatory stress in man, *Circulation*, 41, 485, 1970.
59. **Tuckman, J. and Shillingford, J.,** Effect of different degrees of tilt on cardiac output, and blood pressure in normal man, *Br. Heart J.*, 18, 32, 1966.
60. **Ward, R. J., Danziger, F., Bonica, J. J., et al.,** Cardiovascular effects of change of posture, *J. Aerosp. Med.*, 37, 257, 1966.
61. **Zambrano, S. and Spodick, D. H.,** Comparative responses to orthostatic stress in normal and abnormal subjects, *Chest*, 65, 394, 1974.
62. **Wang, Y., Marshall, R. J., and Shepherd, J. T.,** The effects of changes in posture and of graded exercise on stroke volume in man, *J. Clin. Invest.*, 39, 1051, 1960.
63. **Hainsworth, R. and Al-Shamma, Y. M.,** Cardiovascular responses to upright tilting in healthy subjects, *Clin. Sci.*, 74, 17, 1988.
64. **Smith, J. J., Bonin, M. L., Wiedmeier, V. T., Kalbfleisch, J. H., and McDermott, D. J.,** Cardiorespiratory response of young men to diverse circulatory stresses, *J. Aerosp. Med.*, 45, 583, 1974.
65. **Linde, B. and Hjemdahl, P.,** Effect of tilting on adipose tissue and sympathetic activity in humans, *Am. J. Physiol.*, 242, 161, 1982.

66. **Katkov, V. E., Chestukhin, V. V., and Kakurin, L. I.,** Coronary circulation of healthy man during tilt, *Aviat. Space Environ. Med.,* 56, 741, 1985.

67. **Matalon, S. V. and Farhi, L. E.,** Cardiopulmonary readjustments in passive tilt, *J. Appl. Physiol.,* 47, 503, 1979.

68. **Essandoh, L. K., Duprez, D. P., and Shepherd, J. T.,** Postural cardiovascular reflexes: comparison of responses of forearm and calf resistance vessels, *J. Appl. Physiol.,* 63, 1801, 1987.

69. **Duprez, D. P., Essendoh, L. A., VanHoutte, P. M., and Shepherd, J. T.,** Vascular responses in forearm and calf to contralateral static exercise, *J. Appl. Physiol.,* 66, 669, 1989.

70. **Mengesha, Y. A. and Bell, G. H.,** Forearm and finger blood flow responses to passive body tilt, *J. Appl. Physiol.,* 46, 288, 1979.

71. **Iwase, S., Mano, T., and Saito, M.,** Effects of graded headup tilting on muscle sympathetic activity in man, *Physiologist,* 30 (Suppl.), S62, 1987.

72. **Burke, D., Sundlof, G., and Wallin, B. G.,** Postural effects on muscle nerve sympathetic activity in man, *J. Physiol.,* 272, 399, 1977.

73. **Joyner, M. J., Shepherd, J. T., and Seals, D. R.,** Temporal dissociation of vascular resistance and sympathetic neural discharge in humans, *Physiologist,* 31, 112, 1988.

74. **VonDrischel, H., Fanter, H., Gurtler, H., et al.,** Das Verhalten der Herzfrequenz beim Ubergang vom liegen zum stehen, *Archiv. fur Kreislaufforschung,* 40, 135, 1963.

75. **Brauer, G. and Rossberg, F.,** Zum verhalten des Menschen bei unterschiedlicher Geschwindig keit des Ubergangs vom leigen zur Kopf aufwartsposition, *Acta Biol. Med. Germ.,* 34, 1153, 1975.

76. **Rossberg, F. and Martinez, L.,** Das Ubergangsverhalten der Herzfrequenz in Abhangigkeit von der Atemphase, *Eur. J. Appl. Physiol.,* 50, 291, 1983.

77. **Ewing, D. J., Campbell, I. W., Murray, A., et al.,** Autonomic mechanisms in the initial heart rate response to standing, *J. Appl. Physiol.,* 49, 809, 1980.

78. **Borst, C., Wieling, W., VanBrederode, J. F. M., et al.,** Mechanisms of initial heart rate response to postural change, *Am. J. Physiol.,* H676, 1982.

79. **Wieling, W.,** *Standing, Orthostatic Stress and Autonomic Function in Autonomic Failure,* Bannister, R., Ed., Oxford University Press, New York, 309, 1988.

80. **Mitchell, J. H. and Wildenthal, K.,** Static (isometric) exercise and the heart, *Ann. Rev. Med.,* 25, 368, 1974.

81. **Wieling, W., et al.,** Methods to estimate impairment of cardiac innervation in diabetic neuropathy, *Neth. J. Med.,* 28, 383, 1985.

82. **Smith, J. J., Barney, J. A., Porth, C. J., Groban, L., Stadnicka, A., and Ebert, T. J.,** Transient hemodynamic responses to circulatory stress in normal male subjects of different ages, *Physiologist,* 27, 210, 1984.

83. **Gauer, O. W. and Henry, J. P.,** Neurohormonal control of plasma volume, in *Cardiovascular Physiology II,* Vol. 9, Guyton, A. C. and Cowley, A. W., Eds., University Park Press, Baltimore, 1976, 145.

84. **Hickler, R. B., Wells, R. E., and Tyler, H. R.,** Plasma catecholamine and EEG responses to acute postural change, *Am. J. Med.,* 26, 410, 1959.

85. **Goldstein, D. S., McCarty, R., Polinsky, R. J., and Kopin, I. J.,** Relationship between plasma norepinephrine and sympathetic neural activity, *Hypertension,* 5, 552, 1983.

86. **Esler, M., Jennings, G., Korner, P., Willett I., et al.,** Assessment of human sympathetic nervous system activity from measurements of norepinephrine turnover, *Hypertension,* 11, 3, 1988.

87. **Robertson, D., Johnson, G. A., and Robertson, R. M.,** Comparative assessment of stimuli that release neuronal and adrenal medullary catecholamines in man, *Circulation,* 59, 637, 1979.

88. **Robinson, B. J. and Johnson, R. H.,** Why does vasodilation occur during syncope? *Clin. Sci.,* 74, 347, 1988.

89. **Sander-Jensen, K., Secher, N. H., Astrup, A., et al.,** Hypotension induced by passive head-up tilt: endocrine and circulatory mechanisms, *Am. Physiol. Soc.,* 86, R742, 1986.

90. **Goldsmith, S. R., Francis, G. S., Cowley, A. W., et al.,** Response of vasopressin and norepinephrine to LBNP in humans, *Am. J. Physiol.,* 243, 970, 1982.

91. **Sander-Jensen, K., Mehlsen, J., Stadeager, C., et al.,** Increase in vagal activity during hypotensive lower-body negative pressure in humans, *Am. Physiol. Soc.,* 88, R149, 1988.

92. **Weicker, W.,** Sympathoadrenergic regulation, *Int. J. Sports Med.,* 7 (Suppl.), 16, 1986.

93. **Mark, A. L., Abboud, F. M., and Fitz, A. E.,** Influence of low- and high-pressure baroreceptors on plasma renin activity in humans, *Am. J. Physiol.,* 235, H29, 1978.

94. **Mohanty, P. K., Sowers, J. R., McNamara, C., and Thames, M. D.,** Reflex effects of prolonged cardiopulmonary baroreceptors unloading in humans, *Am. J. Physiol.,* 254, R320, 1988.

95. **Julius, S., Cottier, C., Egan, B., et al.,** Cardiopulmonary mechanoreceptors and renin release in humans, *Fed. Proc.,* 42, 2703, 1983.

96. **Egan, B. M., Julius, S., Cottier, C., et al.,** Role of cardiovascular receptors in the neural regulation of renin release in normal men, *Hypertension,* 5, 779, 1983.

97. **Bannister, R., Sever, P., and Gross, M.,** Cardiovascular reflexes and biochemical responses in progressive autonomic failure, *Brain,* 100, 327, 1977.

98. **Norsk, P. and Epstein, M.,** Effects of water immersion on arginine vasopressin release in humans, *Am. Physiol. Soc.,* 88, 1, 1988.

99. **Rogge, J. D. and Moore, W. W.,** Influence of LBNP on peripheral venous ADH levels in man, *J. Appl. Physiol.,* 26, 134, 1968.

100. **Goldsmith, S. R., Cowley, A. W., Jr., Francis, G. S., and Cohn, J. N.,** Effect of increased intracardiac and arterial pressure on plasma vasopressin in humans, *Am. J. Physiol.,* 246, H647, 1984.

101. **Goldsmith, S. R., Dodge, D., and Cowley, A. W.,** Nonosmotic influences on osmotic stimulation of vasopressin in humans, *Am. Physiol. Soc.,* 87, H85, 1987.

102. **Goldsmith, S. R., Cowley, A. W., Jr., Francis, G. S., and Cohn, J. N.,** Reflex control of osmotically stimulated vasopressin in normal humans, *Am. J. Physiol.,* 248, 660, 1985.

103. **Segar, W. E. and Moore, W. W.,** Regulation of ADH in release in man. *J. Clin. Invest.,* 47, 2143, 1968.

104. **Davies, R. and Forsling, M. L.,** Vasopressin release after postural changes and syncope, *J. Endocrinol.,* 65, 54, 1975.

105. **Baylis, P. H., Stockley, R. A., and Heath, D. A.,** Influences of LBNP upon arginine vasopressin release. *Clin. Endocrinol.,* 9, 89, 1978.

106. **Davies, R., Slater, J. D. H., Forsling, M. L., and Payne, N.,** The response of arginine vasopressin and plasma renin to postural change in normal man, with observation on syncope, *Clin. Sci. Mol. Med.,* 51, 267, 1976.

107. **Ebert, T. J.,** Effect of infusion of vasopressin on hemodynamics and cardiopulmonary baroreceptors, *J. Clin. Invest.,* 77, 1136, 1986.

108. **Zerbe, R., Henny, D., and Robertson, G.,** Vasopressin response to orthostatic hypotension, *Am. J. Med.,* 74, 265, 1983.

109. **Ogihara, T., Shima, J., Hara, H., et al.,** Changes in human plasma atrial natriuretic polypeptide concentration in normal subjects during passive leg raising and whole-body tilting, *Clin. Sci.,* 71, 147, 1986.

110. **Haller, B. G. D., Zust, H., Shaw. S., et al.,** Effects of posture and ageing on circulating atrial natriuretic peptide levels in man, *J. Hypertens.,* 5, 551, 1987.

111. **Cuneo, R. C., Espiner, E. A., Nicholls, M. G., et al.,** Renal, hemodynamic and hormonal responses to atrial natriuretic peptide infusions in normal man, and effect of sodium intake, *J. Clin. Endocrinol. Metab.,* 63, 946, 1986.

112. **Weidmann, P., Hellmueller, B., Uehlinger, D. E., et al.,** Plasma levels and cardiovascular, endocrine, and excretory effects of atrial natriuretic peptide during different sodium intakes in man, *J. Clin. Endocrinol. Metab.,* 62, 1027, 1986.

113. **Ebert, T. J.,** Reflex activation of sympathetic nervous system ANF in humans, *Am. J. Physiol.,* 685, 1988.

114. **Ebert, T. J., Skelton, M. M., and Cowley, A. W.,** Cardiovascular responses to infusions of ANF in humans, *Hypertension,* 11, 537, 1988.

115. **Groban, L., Ebert, T. J., Kreis, D. U., et al.,** Hemodynamic, renal, and hormonal responses to incremental ANF infusion in humans, *Am. Physiol. Soc.,* 89, F780, 1989.

116. **Solomon, L. R., Patterson, J. C., Bobinski, H., et al.,** Effect of posture on plasma immunoreactive ANF concentration in man, *Clin. Sci.,* 71, 229, 1986.

117. **Sakurai, H., Naruse, M. M., Naruse, K., et al.,** Postural suppression of plasma atrial natriuretic polypeptide concentration in man, *Clin. Endocrinol.,* 26, 173, 1987.

118. **Folkow, B.,** Transmural pressure and vascular tone — some aspects of an old controversey, *Arch. Int. Pharmacodyn.,* 139, 455, 1962.

119. **Bedford, T. G. and Dormer, K. J.,** Influence of cerebellar lesions in response to headup tilt in conscious dogs, *Fed. Proc.,* 44, 1728, 1985.

120. **Rothe, C. F.,** The venous system and the physiology of the capacitance vessels in *Handbook of Physiology,* Vol. 3, Sect. 2, Shepherd, J. T. and Abboud, F. M., Eds., American Physiological Society, Bethesda, MD, 1983, 397.

121. **Pickering, T. G., Gribbin, B., Peterson, E. S., et al.,** Effects of exercise and posture on the baroreflex in man, *Cardiovasc. Res.,* 5, 582, 1971.

122. **Frey, M. A. and Doerr, B. M.,** Correlations between ejection times measured from carotid pulse contour and the impedance cardiogram, *Aviat. Space Environ. Med.,* 54, 894, 1983.

123. **Smith, J. J., Muzi, M., Barney, J. A., et al.,** Impedance-derived cardiac indices in supine and upright exercise. *Ann. Biomed. Eng.* 17, 505, 1989.

124. **Frey, M. A. B. and Kenney, R. A.,** Systolic time intervals during combined handgrip and headup tilt, *Aviat. Space Environ. Med.,* 50, 218, 1979.

125. **Spodick, D. H., Meyer, M., and St. Pierre, J. R.,** Effect of upright tilt on the phases of the cardiac cycle in normal subjects, *Cardiovasc. Res.,* 5, 210, 1971.

126. **Hill, D. W. and Merrifield, A. M.,** Left ventricular ejection and the Heather Index during postural changes in man, *Acta Anaesth. Scand.,* 20, 313, 1976.

127. **Weissler, A. M., Leonard, J. J., and Warr, J. V.,** Effects of posture and atropinization on cardiac output, *J. Clin. Invest.,* 36, 1956, 1957.

128. **Eckberg, D. L.,** Parasympathetic cardiovascular control in human disease: a critical review of methods and results, *Am. J. Physiol.,* H581, 1980.

129. **Robinson, B. F., Epstein, S. E., Beiser, G. D., et al.,** Control of heart rate by the autonomic nervous system and interrelations between baroreceptor mechanisms and exercise, *Circ. Res.* 19, 400, 1966.

130. **Zoller, R. P., Mark, A. L., Abboud, F. M., et al.,** The role of low pressure baro-receptors in vasoconstrictor responses in man, *J. Clin. Invest.*, 51, 2967, 1972.

131. **Ebert, T. J., Hughes, C. V., Tristani, F. E., Barney, J. A., and Smith, J. J.,** Effect of age and coronary heart disease on the circulatory responses to graded LBNP, *Cardiovasc. Res.*, 16, 663, 1982.

132. **Grassi, G., Gavazzi, C., Cesura, A. M., et al.,** Changes in plasma catecholamines in response to reflex modulation of sympathetic vasoconstrictor tone by cardiopulmonary receptors, *Clin. Sci.*, 68, 503, 1985.

133. **Eckberg, D. L., Abboud, F. M., and Mark, A. L.,** Modulation of carotid baroreflex responsiveness in man: effects of posture and propranolol, *J. Appl. Physiol.*, 41, 383, 1976.

134. **Loeppky, J. P.,** Cardiorespiratory responses to orthostasis and the effects of propranolol, *Aviat. Space Environ. Med.*, 46, 1164, 1975.

135. **Ferguson, D. W., Thames, M. D., and Mark, A. L.,** Effects of propranolol on reflex vascular responses to orthostatic stress in humans, *Circulation*, 67, 802, 1983.

136. **Tyden, G.,** Aspects of cardiovascular reflex in man, *Acta Physiol. Scand.*, 468, (Suppl.) 7, 1977.

137. **Lindblad, L. E., Atterhog, J. H., and Wallin, B. G.,** Sympathetic activity in muscle nerves, a factor influencing the postural heart rate increase, *Acta Physiol. Scand.*, 111, 509, 1981.

138. **Wolthuis, R. A., Hoffler, G. W., and Johnson, R. L.,** LBNP as an assay technique for orthostatic tolerance, *Aerospace Med.*, 41, 29, 1970.

139. **Mancia, G. and Mark, A. L.,** Arterial and cardiopulmonary reflexes in humans in *Handbook of Physiology*, Vol. 3, Sect. 2, Shepherd, J. T. and Abboud, F. M., Eds., American Physiological Society, Bethesda, MD, 1983, 755.

140. **Beecher, H. K., Field, M. E., and Krogh, A.,** The effects of walking on venous pressure at the ankle, *Scand. Arch. J. Physiol.*, 73, 133, 1936.

141. **Conrad, M. C.,** *Functional Anatomy of the Circulation to the Extremities*, Year Book Medical Publishers, Chicago, 1971.

142. **Rattan, S. N., Glasser, R. M., Servidio, F. J., et al.,** Skeletal muscle pumping via voluntary and electrically induced contractions, *Physiologist*, 28, 363, 1985.

143. **Smith, M. L., Barker, D. J., Cowell, L. L., et al.,** Effects of muscle tension on hemodynamic response to LBNP, *Fed. Proc.*, 44, 816, 1985.

144. **Asmussen, E., Christensen, E. W., and Nielsen, M.,** Puls-Frequenz und korperstelling, *Scand. Arch. Physiol.*, 81, 190, 1939.

145. **Laughlin, M. H.,** Skeletal muscle blood flow: role of muscle pump in exercise hyperemia, *Am. J. Physiol.*, 253, H993, 1987.

146. **Nashner, L. M.,** Fixed patterns of rapid postural responses among leg muscles during stance, *Exp. Brain Res.*, 30, 13, 1977.

147. **Wolthuis, R. A., Bergman, S. A., and Nicogossian, A. E.,** Physiological effects of locally applied reduced pressure in man, *Physiol. Rev.*, 54, 566, 1974.

148. **Brown, E., Goei, J. S., Greenfield, A. D. M., et al.,** Circulatory responses to stimulated gravitational shifts of blood in man, *J. Physiol.*, 183, 607, 1966.

149. **Musgrave, F. S., Zechman, F. W., and Mains, R. C.,** Changes in total leg volume during LBNP, *Aerospace Med.*, 40, 602, 1969.

150. **Ebert, T. J., Stowe, D. F., Barney, J. A., Kalbfleisch, J. H., and Smith, J. J.,** Summated circulatory responses of thermal and baroreflexes in man, *J. Appl. Physiol.*, 52, 184, 1982.

151. **Milledge, R. D. and Zechman, F. W.,** Radioisotope studies of the chest during LBNP, *Radiology*, 90, 654, 1968.

152. **Zechman, F. W., Musgrave, F. S., Mains, R. C., et al.,** Respiratory mechanics and pulmonary diffusing capacity with LBNP, *J. Appl. Physiol.* 22, 247, 1967.

153. **Menninger, R. P., Mains, R. G., Zechman, F. W., et al.,** Effect of two weeks bed rest on venous pooling in the lower limbs, *Aerospace Med.*, 40, 1323, 1969.

154. **Smith, M. L., Graitzer, H. M., Hudson, D. L., and Raven, P. B.,** Baroreflex function in endurance and static exercise, *J. Appl. Physiol.*, 64, 585, 1988.

155. **Raven, P. B., Rhom-Young, D., and Blomqvist, C. G.,** Physical fitness and cardiovascular response to LBNP, *J. Appl. Physiol.*, 56, 138, 1984.

156. **Convertino, V. A., Sather, T. M., Goldwater, D. J., et al.,** Aerobic fitness does not contribute to prediction of orthostatic intolerance, *Med. Sci. Sports Exerc.*, 18, 551, 1986.

157. **Montgomery, L. D., Kijk, P. J., and Payne, P. A.,** CV responses of men and women to LBNP, *Aviat. Space Environ. Med.*, 48, 138, 1977.

158. **Mack, G. W., Shi, X., Nose, H., Tripathi, A., and Nadel, E. R.,** Diminished baroreflex control of forearm vascular resistance in humans, *J. Appl. Physiol.*, 63, 105, 1987.

159. **Takeshita, A., Jingu, S., Imaizumi, T., et al.,** Augmented cardiopulmonary baroreflex control of forearm vascular resistance in young athletes, *Circ. Res.*, 59, 43, 1986.

160. **Stevens, C. M. and Lamb, L. E.,** Effects of lower body negative pressure on the cardiovascular system, *Am. J. Cardiol.*, 16, 506, 1965.

161. **Nixon, J. V., Murray, R. G., Leonard, P. D., Mitchell, J. H., and Blomqvist, C. G.,** Effects of large variations in preload: left ventricular performance characteristics in normal subjects, *Circulation,*, 65, 698, 1982.

162. **Nutter, D. O., Hurst, V. W., and Murray, R. H.,** Ventricular performance during graded hypovolemia induced by LBNP, *J. Appl. Physiol.,* 26, 23, 1969.

163. **Ahmad, M., Blomqvist, C. G., Mullins, C. B., et al.,** Left ventricular function during LBNP, *Aviat. Space Environ. Med.,* 48, 512, 1977.

164. **Tomaselli, C. M., Frey, M. A. B., and Kenney, R.A.,** Hysteresis in response to descending and ascending LBNP, *J. Appl. Physiol.,* 63, 710, 1987.

165. **Murray, R. H., et al.,** Hemodynamic effects of graded hypovolemia and vasodepressor syncope induced by LBNP, *Am. Heart J.,* 76, 799, 1968.

166. **Tripathi, A. and Nadel, E. R.,** Forearm, skin, and muscle vasoconstriction during LBNP, *J. Appl. Physiol.,* 60, 1535, 1986.

167. **Rea, R. F. and Wallin, B. G.,** Sympathetic outflow to arm and leg during LBNP, *FASEB J.,* 2, A313, 1988.

168. **Victor, R. G. and Leimback, W. M.,** Effects of LBNP on sympathetic discharge to leg muscles in the human, *J. Appl. Physiol.,* 63, 2558, 1987.

169. **Anderson, E., Sinkey, C. A., and Mark, A. L.,** Regulation of large peripheral artery during LBNP, *FASEB J.,* 2, A313, 1988.

170. **Ardill, B. L., Bannister, R. G., Fenten, P. H., et al.,** Circulatory responses of the lower body to subatmospheric pressure, *J. Physiol.,* 193, 57, 1967.

171. **Hirsch, A. T., Levenson, D. J., Cutter, S. S., et al.,** Regional vascular responses to prolonged LBNP in normal subjects. *Am. J. Physiol.,* 257, H219, 1989.

172. **Graybiel, A. and McFarland, R. A.,** Use of tilt table test in aviation medicine, *J. Aviat. Med.,* 11, 194, 1941.

173. **Brigden, W., Howarth, S., and Sharpey-Schafer, E. P.,** Postural changes and peripheral blood-flow during vasovagal fainting reactions as a result of tilting, *Clin. Sci.,* 9, 75, 1950.

174. **Newberry, P. D., Hatch, A. W., and McDonald, J. M.,** Cardiorespiratory events preceding syncope induced by LBNP and head up tilt, *Aerospace Med.,* 41, 373, 1970.

175. **Stevens, P. M.,** Cardiovascular dynamics during orthostatics and the influence of intravascular instrumentation, *Am. J. Cardiol.,* 17, 211, 1966.

176. **Lind, A. R., Leithead, C. S., and Menicol, G. N.,** Cardiovascular changes during syncope induced by tilting men in the heat, *J. Appl. Physiol.,* 25, 268, 1968.

177. **McHenry, L. C., Fazekas, J. F., and Sullivan, J. F.,** *Central Hemodynamics of Syncope,* Yohr, M. D., Ed., Springer Verlag, New York, 1960, 106.

178. **Engel, G. L., Ferris, E. B., and Logan, M.,** Hyperventilation: analysis of clinical symptomology, *Ann. Int. Med.,* 27, 683, 1947.

179. **McHenry, L. C., Fazekas, J. F., and Sullivan, J. F.,** Cerebral hemodynamics of Syncope, *Am. J. Med. Sci.,* 241, 173, 1961.

180. **Taylor, N. B. and Noble, R. N.,** Hormones in hypovolemia, *Proc. Soc. Exp. Biol.,* 73, 207, 1950.

181. **Sunahara, F., Dunemann, D., and Edholm, O.,** Effect of posterior pituitary extract in man, *Fed. Proc.,* 8, 152, 1949.

182. **Verbalis, J. G., Richardson, D. W., and Stricker, E. M.,** Vasopressin release in response to nausea producing agents and cholecystokinin in monkeys, *Am. J. Physiol.,* 252, 749, 1987.

183. **Elliot, R. S.,** *Stress and the Heart,* Futura Publishing, Mt. Kisco, NY, 1974.

184. **Callahan, R., Edelman, E. B., Smith, M. S., and Smith, J. J.,** Study of incidence and characteristics of blood donor "reactors", *Transfusion,* 3, 76, 1963.

185. **Epstein, S. E., Stampick, M., and Beiser, G. D.,** Role of the capacitance and resistance vessels in vasovagal syncope, *Circ. Res.,* 37, 524, 1968.

186. **Wallin, B.G. and Sundloff, G.,** Sympathetic outflow to muscles during vasovagal syncope, *J. Auton. Nerv. Syst.,* 6, 287, 1982.

187. **Uvnas, B.,** Sympathetic vasodilator system and blood flow, *Physiol. Rev.,* 40 (Suppl. 4) 69, 1960.

188. **Löfving, B.,** Cardiovascular adjustment induced from the rostral cingulate gyrus, *Acta Physiol. Scand.,* 53 (Suppl. 184), 1, 1961.

189. **Engel, G. L.,** *Fainting,* Charles C. Thomas, Springfield, IL, 1982.

190. **Murray, R. H., Bowls, J. A., and Goltra, E. R.,** Comparison of footboard and saddle supports for orthostatic tilt table tests, *J. Appl. Physiol.,* 21, 1409, 1966.

191. **Spodick, H. and Lance, V. Q.,** Comparative orthostatic responses, standing versus headup tilt, *Aviat. Space Environ. Med.,* 48, 432, 1977.

192. **Hyatt, K. H., Jacobsen, L. G., and Schneider, V. S.,** Comparison of 70° tilt, LBNP and passive standing as measures of orthostatic tolerance, *Aviat. Space Environ. Med.,* 46, 801, 1975.

193. **Musgrave, F. S., Zechman, F. W., and Mains, R. C.,** Comparison of the effects of 70° tilt and LBNP in man, *Aerospace Med.,* 42, 1065, 1971.

194. **Schvartz, E.,** Prediction and management of orthostatic insufficiency, *Military Med.,* 138, 222, 1973.

195. **Aschoff, J. and Aschoff, J.,** Tages periodik der orthostatischen Kreislaufreaction, *Pflügers Arch,*, 306, 146, 1969.

196. **Lipsitz, L. A. and Fullerton, K. J.,** Post-prandial blood pressure reduction in healthy elderly, *J. Am. Geriatr. Soc.*, 34, 267, 1986.

197. **Myers, H. A. and Honig, C. R.,** Influence of initial resistance on magnitude of response to vasomotor stimuli, *Am. J. Physiol.,* 216, 1429, 1969.

198. **Hatch, R. C., Hughes, R. W., and Bozivich, H.,** Effect of resting blood pressure and pressor response to drugs and carotid occlusion, *Am. J. Physiol.,* 213, 1515, 1967.

199. **Schvartz, E. and Meyerstein, N.,** Relation of tilt tolerance to aerobic capacity and physical characteristics., *Aerospace Med.,* 43, 278, 1972.

200. **Wolthuis, R. A., Hull, D. H., Fischer, J. R., et al.,** Blood pressure variability in orthostatic testing, *Aviat. Space Environ. Med.,* 50, 774, 1979.

201. **Schvartz, E.,** Reliability of quantitative tilt table data, *Aerospace Med.,* 39, 1096, 1968.

202. **Xiangchang, Z., Yaming, F., QiuLu, X., et al.,** Relationship between cardiovascular tolerance before and after bedrest, *Physiologist,* 31, S102, 1988.

203. **Vogt, F. B.,** Objective approach to the analysis of tilt table data, *Aerospace Med.,* 37, 1195, 1966.

204. **Benjamin, F. B., Townsend, J. C., Vinograd, S. P., et al.,** Integrated scoring of tilt table response, *Aerospace Med.,* 39, 159, 1968.

205. **Berry, C. A.,** Summary of medical experiences in the Apollo 7 through 11 manned spaceflights, *Aerospace Med.,* 41, 500, 1970.

Chapter 2

PHYSICAL FITNESS AND ORTHOSTATIC TOLERANCE

Thomas J. Ebert and Jill A. Barney

TABLE OF CONTENTS

I. INTRODUCTION

The cardiovascular responses to orthostatic (gravitational) stress have been intensely studied over the past century. A few of these studies have been directed toward ascertaining the influence of athletic conditioning on tolerance to orthostatic stress. An early perception was that athletes, because of better cardiovascular conditioning, had improved circulatory responses to orthostatic stresses. For example, research findings in the 1920s suggested that athletes made better circulatory adjustments upon standing than did nonathletes.[1,2] Supporting data was provided several decades later by Shvartz et al.,[3] who studied the responses of 18 healthy subjects during head-up tilt-table testing. Syncope occurred in four subjects. Retrospective analyses revealed that those who experienced syncope could perform fewer sit-ups than the 14 subjects who tolerated the tilt stress. These investigators concluded that conditioning of the abdominal muscles was important for maintaining orthostatic tolerance. In later work by these investigators,[4] orthostatic stress, which consisted of quiet standing, was employed to study the relationship between aerobic capacity, determined by peak oxygen uptake during treadmill exercise, and the incidence of syncope in 28 men. They noted that the frequency of fainting was less in 12 of these men, and they had higher average aerobic capacities than the 16 men who were less tolerant to the stress.

The early belief that cardiovascular fitness was associated with improved orthostatic tolerance was questioned as early as 1940 by several investigators who compared athletes and nonathletes, and found that orthostatic tolerance did not differ.[5,6] Since these early reports, a considerable amount of research has been focused on determining whether the degree of aerobic or physical fitness has any consistent influence on the regulation of blood pressure during orthostatic stress. Much of the interest and research on this topic has originated through the National Aeronautics and Space Administration and the Air Force, where the tolerance to large head-to-foot acceleration forces ($+G_z$) during launch and reentry from space and during high-speed maneuvering in sophisticated jet fighter planes has been a long-standing concern. This chapter will review and summarize much of the research that deals with the issue of aerobic fitness and its effect or lack of effect on the regulation of blood pressure during orthostatic stress. Convertino has pointed out in his recent review on aerobic fitness, endurance training and orthostatic intolerance[7] that the published experimental data that demonstrate an adverse effect of aerobic fitness on orthostatic tolerance is primarily based upon only a few cross-sectional studies in which syncope occurred in less than 25 highly trained competitive runners. The majority of the published research findings, which examined nearly 300 subjects, have failed to support these observations. In the first half of this chapter, we will review these research findings in a systematic fashion and attempt to ascertain why divergent results exist.

The ability to tolerate orthostatic stress is dependent on a series of interrelated factors, which include the pumping ability or "cardiac reserve" of the heart, the intravascular volume status and the distribution of this volume during orthostatic stress, the inherent compliance of veins in the pelvis and lower extremities, which accommodate volume during orthostatic stress, and the integrity of cardiovascular reflexes that constrict arteriolar and venous smooth muscle and elicit secretion of vasoactive hormones. We have focused the latter portions of this chapter on research that has contrasted the aerobically fit to sedentary individual in terms of these specific physical, neural, and endocrine factors, which are known to be involved in the cardiovascular responses to orthostatic stress.

II. RESEARCH FINDINGS WHICH SUGGEST THAT ORTHOSTATIC TOLERANCE MAY NOT BE INFLUENCED BY AEROBIC FITNESS

With the establishment of a reliable standard for assessing physical fitness, i.e., maximum oxygen uptake, a number of studies carried out over the last 20 years have sought a significant statistical relationship between endurance capacity (Vo_{2max}) and tolerance to orthostatic stress. Klein et al.[8] reported in 1969 on a group of 12 athletes who, due to many years of intense physical training (long-distance running, bicycling, or ice skating), had about a 50% higher maximum work capacity than that of 12 healthy students (Vo_{2max}, 65 versus 44 ml/kg/min). The orthostatic tolerance to both a 20-min, 90° head-up tilt and centrifugal ($+G_z$) forces was found to be identical between groups. Subsequent findings in a separate study by this group supported their initial conclusion and extended this by demonstrating that when athletes faint, they do so at a lower heart rate than nonathletes who faint.[9]

Results from a cross-sectional study carried out on eight endurance-trained female runners (Vo_{2max} 56.8 ml/kg/min) and eight sedentary (39.4 ml/kg/min) young female volunteers undergoing progressive lower body negative pressure (LBNP) to –50 mmHg have been reported by Hudson et al.[10] Syncopal episodes did not occur in either group during LBNP, and cardiovascular responses (heart rate, blood pressure, cardiac index, forearm vascular resistance, and calf circumference increases) to LBNP were similar in the two groups. They summarized their findings by stating that their data "failed to conclusively demonstrate a fitness-related difference in cardiovascular responses to LBNP in females".

A somewhat different approach was undertaken by Sather and colleagues,[11] who determined the tolerance of 18 men, ages 29 to 51 years, to progressive LBNP, which was terminated upon the onset of presyncopal signs (sudden hypotension or bradycardia) or symptoms (dizziness, nausea, discomfort). Subjects were then classified into high-tolerance (n = 10) or low tolerance (n = 8) groups. There were no significant differences found between the two groups when age, height, weight, percent body fat, total blood volume, and Vo_{2max} were retrospectively compared. Greater tolerance to LBNP was, however, associated with larger increases in heart rate and peripheral resistance, i.e., greater baroreflex-mediated compensatory responses to controlled central hypovolemia.

Frey and colleagues used a somewhat similar protocol in a larger study on 45 healthy women.[12] These women had peak aerobic capacities ranging from 23 to 55.3 ml/kg/min. Each underwent graded LBNP to –50 mmHg. There were 6 women who developed signs and symptoms of impending syncope, but retrospective analyses indicated that their average Vo_{2max} values did not differ from those of the 39 tolerant females. These authors concluded that orthostatic tolerance was not related to aerobic capacity.

Convertino et al.[13] applied multiple regression statistics to their previously reported[11] data collected from 18 subjects to derive a predictive model for LBNP tolerance. These healthy subjects (aged 29 to 51 years) had been stressed to presyncope with graded progressive LBNP. An index of LBNP tolerance was derived by summing the products of negative pressure and time (in minutes) of each test stage completed prior to termination. They included numerous physical and physiological variables, including age, height, weight, Vo_{2max}, body fat, blood volume, arterial pulse volume, thigh fluid accumulation index, thigh compliance index, heart rate, stroke volume, cardiac output, mean arterial pressure, and peripheral resistance in their analyses. The type of exercise (weight lifting, biking, running, swimming, etc.) was not controlled for in this study. Routine linear regression statistics were applied to seek a relationship of these variables to Vo_{2max}. Significant inverse correlations were derived between Vo_{2max} and age (r = -0.55), weight (r = -0.56), and body fat (r = -0.94), and significant direct correlations were obtained when relating Vo_{2max} to blood volume (r = 0.81) and thigh compliance index (r = 0.50). Vo_{2max} did not correlate highly or significantly with the peak level of LBNP achieved prior to the onset of presyncopal symptoms. A stepwise regression statistic was applied to determine the best predictive model for LBNP tolerance. This was achieved by including the thigh fluid accumulation index, blood volume, and thigh compliance index in the regression model. The accumulation of fluid in the thigh was the major contributor (82%) to the prediction model. Based upon this model, the more accumulation of fluid in the thigh, the greater the predicted tolerance to LBNP. Although neither the thigh compliance index nor the blood volume were significantly correlated with LBNP tolerance, each provided a significant negative contribution to the LBNP tolerance prediction model. Thus, lower blood volume and a reduced compliance of the thigh would predict a high tolerance to LBNP. Several seemingly incongruent results arise from this statistical analysis. First, the capacity to accommodate a large amount of blood pooling in the thigh is apparently not related to the compliance of the thigh veins. Secondly, in this study, a large blood volume predicts a low tolerance to LBNP and, although an increase in the central blood volume has been reported in endurance-trained individuals,[14] Vo_{2max} did not contribute to the prediction model for LBNP tolerance.

Several different statistical modeling procedures were employed by Ludwig and colleagues[15] in a retrospective analysis of 25 women and 22 men who had undergone centrifuge studies conducted at NASA/Ames Research Center in Moffett Field, CA. Supine bicycle ergometry was used to determine Vo_{2max}, which ranged from 18 to 50 ml/kg/min. Two biostatistical modeling procedures, proportional hazard and logistic discriminant function, were used to estimate the risk of greyout during centrifugation (chosen as the dependent variable). Study variables included gender, age, weight, height, percent body fat, baseline heart rate and blood pressure, Vo_{2max} and plasma volume. Only age, gender, height, and plasma volume significantly contributed to the risk equation. Tall, young women with large plasma volumes were at increased risk of orthostatic intolerance. They also concluded that tolerance to $+G_z$ acceleration was independent of aerobic fitness.

One of the first longitudinal studies to carefully investigate orthostatic tolerance before and after an exercise training program that consisted entirely of running was carried out by Cooper et al.[16] Orthostatic tolerance to $+G_z$ forces produced by centrifugation were determined in 11 subjects who were subsequently randomized into 2 groups: 6 exercisers and 5 controls. The exercising group engaged in a daily progressive running program for 3 months while the controls were asked to avoid vigorous exercise. At the conclusion of the 3 months, the exercising group had increased their endurance capacity by 13%. Despite this, there were no significant differences noted between the two groups in their ability to tolerate $+G_z$ forces.

More recent longitudinal studies by Sheldahl et al.[17] reported the effects of 6 months of endurance exercise training (running) on 12 middle-age men who had undergone pretraining head-up tilt testing. Vo_{2max} increased from 33 to 41 ml/kg/min after training. They reported that baseline heart rate was decreased and stroke volume was increased by the training program. However, cardiovascular responses and orthostatic tolerance to short-term postural stress were not altered by training.

Additional longitudinal studies determined orthostatic tolerance to head-up tilt testing before and after an 8-d cycle ergometer training program, which increased Vo_{2max} by 8.3%,[18] and before and after a 6-month exercise program consisting of calisthenics, rope jumping, stationary cycling, and weight lifting, which increased Vo_{2max} by 18%.[19] Neither study demonstrated an adverse effect of "nonrunning" exercise training on orthostatic tolerance.

In a recent unpublished study,[20] eight young male students underwent a 10-week aerobic exercise training protocol that consisted of alternating sessions on a treadmill and cycle ergometer. This program resulted in a 21% increase in Vo_{2max} (45.7 to 55.2 ml/kg/min). Subjects underwent a presyncopal-limited LBNP test before and after training, which consisted of progressive, 10 mmHg increases in LBNP until one of the following criteria was met: (1) a SBP < 70 mmHg or (2) a decrease of SBP of > 25 mmHg, DBP of > 15 mmHg, or HR of > 15 bpm in two consecutive 1-min readings. The increases in calf circumference and heart rate, and the decreases in forearm blood flow and systolic pressure, produced by LBNP were not different after training. Three indices of LBNP tolerance were derived: (1) duration of exposure to LBNP, (2) magnitude of negative pressure at onset of presyncope, and (3) cumulative stress, determined by summing the products of the duration and pressure magnitude of each LBNP level. None of these indices of tolerance were altered following the 10-week training program. These authors concluded that there was no discernable adverse effect of improved aerobic capacity on orthostatic tolerance over a 10-week training period.

To summarize these cross-sectional and longitudinal studies, it appears that general aerobic fitness does not contribute to orthostatic intolerance. The longitudinal studies suggest that training programs that lasted from 8 d to 6 months and improved aerobic capacity to varying degrees did not adversely influence orthostatic tolerance. The two studies that employed elaborate statistical analyses to groups of varying fitness conform in their conclusion that increased intravascular volume is associated with poor tolerance to orthostatic stress, but neither study was able to relate Vo_{2max} to orthostatic tolerance. In most cross-sectional studies, a good age match was achieved between fit and unfit groups; however, in each aerobically fit group, the type of training to achieve aerobic fitness was not controlled for. High aerobic capacity was achieved in fit groups by running, cycling, skating, or the method was not reported in these cross-sectional studies. In only one report was the aerobically fit group trained entirely by one type of exercise (running), and in this report, LBNP was applied to only 50 mmHg, which did not produce presyncopal signs or symptoms.[10]

In the next section, we will examine studies focused primarily on the endurance-trained *runner* who has been involved in competitive running for many years. These individuals appear to be less tolerant of orthostatic stress than the aerobically fit group, who endurance train with other types of exercise.

III. RESEARCH FINDINGS WHICH SUGGEST THAT ORTHOSTATIC TOLERANCE MAY BE INFLUENCED BY AEROBIC FITNESS

Much of the evidence that will be presented to demonstrate the existence of orthostatic intolerance in the aerobically trained athlete has been derived from cross-sectional studies on a relatively small number of subjects, and most of the data have been published only in abstract form or in NASA proceedings. Some of these reports lack sufficient information to allow critical evaluation of the research. Nevertheless, these reports have stimulated a considerable amount of interest in the topic of orthostatic tolerance and endurance training. Part of this interest was fueled by a published report by Stegemann's group in 1974,[21] which suggested that carotid baroreflex control of blood pressure was diminished in trained athletes (see later discussion of baroreceptors). The initial intriguing suggestion that aerobic conditioning *reduced* orthostatic tolerance was made by Greenleaf et al. in 1975.[22] Seven men were exposed to graded gravitational stress induced by centrifugal acceleration. Subjects were stressed to an endpoint determined by loss of central vision (blackout). The study began with a 14-d ambulatory control period followed by three 14-d bedrest periods, each of which were separated by a 21-d recovery period. During two of the three bedrest periods, supine subjects performed one of two types of exercise, isometric or isotonic (cycle ergometry), for 30 min at 68% of Vo_{2max}. The exercise was performed twice daily. Orthostatic tolerance was reduced when bedrest was undertaken without exercise, while tolerance was improved following bedrest with 14 d of isometric exercise. However, orthostatic tolerance was reduced following 14 d of bicycle exercise during bedrest.

Several observations were made in the late 1970s by Luft and colleagues, which supported the suggestion that orthostatic tolerance might be reduced in aerobically trained individuals.[23,24] They performed a series of ground-based studies for NASA and reported that the average LBNP tolerance was 42% less in five *runners* (mean cycle ergometer Vo_{2max} = 50 ml/kg/min) than in five nonrunners (Vo_{2max} = 34 ml/kg/min). They also reported that runners showed a greater propensity for displacement of blood volume into their lower extremities during LBNP, which suggested that venous compliance was greater in the five fit individuals. In a separate publication, Klein and co-workers[25] used the data from Luft et al.[23] to derive a significant inverse relationship (r = -0.60) between Vo_{2max} and LBNP tolerance, and a significant direct relationship (r = 0.72) between Vo_{2max} and lower limb compliance. Research findings from a second study by Luft and colleagues supported their earlier findings.[24] Forty-seven subjects were studied, of which 13 were endurance *runners*, 12 were weight lifters, 12 were swimmers, and 10 were sedentary individuals. Their average measured maximum aerobic capacities were 51, 39, 35, and 35 ml/kg/min, respectively. The average tolerance to progressive LBNP of the *runners* was less than that of the nonrunners. Moreover, the runners were noted to have larger changes in leg (calf) volume and smaller heart rate increases during LBNP compared with the pooled group of nonrunners. They concluded that exercise training that leads to high aerobic capacity reduces the tolerance to orthostatic stress produced by LBNP. They proposed two potential mechanisms to account for their results: an increase in leg venous compliance in runners and an increase in cardiac parasympathetic tone, which limits cardioacceleration during orthostasis in runners with high aerobic capacities.

Independent research by Mangseth and colleagues[26] provided additional preliminary data to support the findings of Luft and colleagues. In this study, four aerobically trained subjects (average Vo_{2max} of 66.8 ml/kg/min) fainted during head-up tilt, while four control subjects (average Vo_{2max} of 52 ml/kg/min) tolerated 30 minutes of tilt with no signs of syncope. The type of exercise used to achieve aerobic fitness was not reported. In contrast to the findings of Luft et al.,[23] these investigators noted that syncope was not related to an inordinate amount of caudal blood pooling but was more closely related to a fundamental deficiency in the regulation of peripheral resistance in the four aerobically trained subjects.

Several recent published research findings supporting the hypothesis that aerobic fitness diminishes orthostatic tolerance have been provided by Raven, et al.[27,28] In these cross-sectional studies, average-fit (Vo_{2max} = 38 to 42 ml/kg/min) and high-fit individuals (Vo_{2max} = 62 to 70 ml/kg/min) were exposed to progressive increases in LBNP to 50 mmHg. This stress did not produce syncope; however, careful assessment of the cardiovascular responses to each increment of LBNP revealed that the high-fit individuals, who consisted primarily of elite distance runners, had larger decreases in systolic blood pressure than the average-fit subjects. This hypotension could be explained by two deficient responses: less tachycardia and diminished peripheral vasoconstriction in the high-fit subjects. Peripheral venous pooling (determined by calf-circumference changes during LBNP) was not different between groups. Unfortunately, as in most previous studies, the precise degree of reduction in thoracic blood volume was not determined and, therefore, comparison of responses between groups was based solely upon the absolute level of LBNP.

The only longitudinal study to demonstrate that orthostatic tolerance may be adversely affected by exercise programs that emphasize aerobic fitness has been recently reported by Pawelczyk et al.[29] Cardiovascular responses to a 10-min, 70° head-up tilt were compared before and after a 7-week exercise program which consisted of running, rope skipping, and aerobic dance. Average initial Vo_{2max} was 41 ml/kg/min and increased subtly to 45 ml/kg/min. Impedance plethysmography was used to assess calf-volume increases produced by venous occlusion of the thigh. These increases were significantly greater after the training program. Moreover, during tilt, systolic pressure was not maintained as well after training.

There seems to be a fundamental difference in orthostatic tolerance in individuals who undergo strength training compared with endurance-type training, which produces cardio-respiratory changes to improve aerobic capacity. For example, Epperson et al.[30] studied 24 young men who were divided into three groups: controls (no training) and two groups who underwent a 12-week physical training program, which consisted of running or weight training. There was no change in the $+G_z$ tolerance of the runners and controls; however, the weight trainers increased their $+G_z$ tolerance significantly. It has been suggested that individuals who undergo strength training may reduce the amount of caudal venous blood pooling during LBNP compared with those who are endurance trained,[28] and this may be related to a larger muscle mass of the lower extremities, which restricts venous pooling.[31] Supporting data from cross-sectional studies conducted by Smith and colleagues[32] suggest that this increased tolerance in weight-trained individuals may be due to efficient baroreceptor reflexes, which maintain their ability to regulate heart rate and peripheral resistance during LBNP stress.

IV. EFFECTS OF AEROBIC FITNESS ON PHYSICAL, NEURAL, AND HORMONAL FACTORS INVOLVED IN BLOOD PRESSURE HOMEOSTASIS DURING ORTHOSTATIC STRESS

In the previous sections of this chapter, research findings either opposing or supporting the hypothesis that aerobic fitness reduces orthostatic tolerance have been reviewed. In this section, research that has focused on quantifiable differences between fit and unfit individuals in terms of anatomical or pathological changes in the cardiovascular system, specific neural reflexes, and vasoactive hormone production, which are known to be involved in blood pressure regulation, will be examined. These response comparisons may improve our understanding of orthostatic tolerance and its relationship to endurance training.

A. ENDURANCE-TRAINING-INDUCED CARDIOVASCULAR CHANGES THAT MAY BEAR IMPORTANTLY ON ORTHOSTATIC TOLERANCE

Endurance-trained athletes have larger total blood volumes per kilogram body weight compared to untrained individuals.[14,33] A direct relationship has been drawn between work capacity and blood volume.[34] This increased blood volume is due primarily to an increase in plasma volume.[35] Because orthostatic intolerance is clearly related to states of reduced blood volume such as dehydration or hemorrhage, it would seem logical to conclude that the "hypervolemia" of the endurance-trained athlete may be protective by maintaining cardiac preload, cardiac output, and blood pressure during orthostatic stress. However, several other related factors may occur with a chronic increase in blood volume. For example, low-pressure baroreceptor afferent activity may be initially augmented, which would elicit reflex reductions in peripheral sympathetic outflow, plasma renin activity (PRA), and arginine vasopressin (AVP), however, these receptors undoubtedly reset with time[36] and could conceivably become less sensitive. For example, further loading of these receptors by water immersion to the neck results in a diuresis and natriuresis that is due to both atrial-renal reflexes, which reduce renal sympathetic tone, and reductions in PRA, aldosterone, and AVP.[37,38] Research findings indicate that this diuresis and natriuresis is less in athletes compared with sedentary individuals.[39]

Moreover, a chronic increase in blood volume may also alter the distensibility of the vascular smooth muscle, especially in the large reservoir compartments on the venous side of the circulation.

Another well-known cardiovascular adaptation to endurance training is a reduction in the resting heart rate. This may be due to an enhanced resting cardiac vagal tone,[40-43] a withdrawal of cardiac sympathetic activity,[44,45] or nonautonomic adaptations, such as muscarinic or adrenergic receptor changes, which occur at the sinoatrial node.[46-48] There are research data indicating that cardioacceleration in response to exercise[49] and orthostatic stress[27,28] is reduced in the endurance-trained individual. This may be due to a reduced sympathetic drive or an enhanced vagal restraint during these stresses.

Postganglionic sympathetic efferent activity directed to the smooth muscle vasculature of the skeletal muscle in the leg is not different in trained cyclists and untrained subjects.[50] Moreover, resting sympathetic activity is not altered by an 8-week training program. However, it is unknown how much neurotransmitter is released at the neuromuscular junction per endplate depolarization, nor is it certain as to how sensitive the postjunctional sites are to neurotransmitters in the endurance-trained individual compared with sedentary subjects.[7,49]

B. LEG COMPLIANCE AND ITS CONTRIBUTION TO ORTHOSTATIC TOLERANCE

A considerable amount of research that has examined orthostatic tolerance in the aerobically fit and less-fit individual has included some form of measurement on the lower extremity to assess the degree of caudal displacement of blood volume. If the compliance of the veins in the lower extremities is increased in the aerobically fit individual, then conceivably more blood could be accommodated during orthostatic stress, which might lead to a greater reduction in venous return, stroke volume, and cardiac output, eliciting syncope. The most common procedure to measure compliance has been to place a Hg-in-silastic double-stranded strain gauge around the calf to record calf circumference increases during LBNP or venous occlusion of the thigh.[10,12,23,26-28,32] Unfortunately, one limitation of this technique is that differences observed between fit and sedentary subjects could be due to several factors, such as altered venous compliance of the calf, different rates of extravasation of fluids across the capillaries and venules, or a different number of capacitance vessels in the calf. Another limitation of the strain-gauge method is that only circumference changes that occur immediately below the gauge are actually measured. From this, mathematical calculations are used to estimate the total calf volume changes. Perhaps a more accurate measurement can be obtained with impedance plethysmography. This has been done with a tetrapolar impedance system in which the outer two electrodes transmit a current into a leg segment and the inner two electrodes detect changes in impedance.[11,13,18,29] When this is applied to the lower leg, impedance changes during LBNP, tilt, or venous occlusion of the thigh must be due to blood volume changes in the leg, since other impedance-sensitive tissues, e.g., bone, fat, and muscle, do not change. However, one must be certain that electrode movement does not occur or the readings may be errant.

In a NASA publication, Luft et al.[23] found an increased incidence of syncope in five runners compared with five sedentary volunteers during LBNP stress. Calf strain-gauge measurements were used to derive the rate of calf volume change per unit of LBNP stress (the product of time of exposure and level of negative pressure). This index was significantly higher in the runners compared with the nonrunners. Klein et al.[25] applied linear regression statistics to Luft's data and demonstrated a direct relationship between Vo_{2max} and calf compliance in the ten volunteers (r = 0.72). Two supporting studies comparing aerobically fit men[28] and women[10] to sedentary individuals of the same age and gender employed the calf strain gauge to assess calf compliance during LBNP stress. In both studies, the percentage increase in calf circumference and/or volume was greater in the more fit individuals. Several longitudinal studies suggest that calf pooling of blood during orthostatic stress may be increased following an exercise training program. Convertino et al.[18] reported increases in the amount of fluid pooled in the legs after an 8-d bicycle training, but this did not adversely influence tilt tolerance. Pawelczyk et al.[29] used calf impedance measurements to ascertain the amount of blood sequestered during graded venous occlusion of the thigh, before and after a 7-week aerobic exercise program. They noted larger calf volume increases after training in their ten subjects.

Several studies have been unable to show differences between aerobically fit and unfit individuals.[26,27,51] Raven et al.[27,51] calculated calf volume changes using venous occlusion of the thigh and LBNP (separately) in fit (Vo_{2max} 70 to 72 ml/kg/min) and unfit (Vo_{2max} = 41 ml/kg/min) subjects. They noted that the initial calf circumference and calculated baseline calf volumes were less in the fit subjects. Moreover, absolute increases in calf volume were similar between groups.

Convertino et al.[31] used computed tomography scans of the lower leg of ten healthy males to determine the cross-sectional areas of muscle, bone, and fat. Calf compliance was measured during repeated venous occlusions of the thigh with a Hg-in-silastic strain gauge. They found a significant inverse relationship between both total and calf-muscle cross-sectional area and calf compliance, but no relationship between Vo_{2max} (range = 30 to 53 ml/kg/min) and calf compliance. They conjectured that endurance-trained runners may have less muscle mass of the calf than endurance-trained nonrunners, who use more resistive-type calf exercises, e.g., cycling, hopping, or stair climbing. A smaller calf muscle mass might reduce the tendency for muscles to resist venous engorgement, leading to an increased venous pooling during orthostatic stress in endurance-trained *runners*.

C. ARTERIAL BAROREFLEX CONTROL OF CARDIAC FUNCTION

The primary acute mediators of blood pressure homeostasis are the arterial baroreceptor reflexes. Baroreceptors are located in the aortic arch and carotid sinus regions, and respond to changes in transmural pressure. The efferent component of this reflex consists primarily of changes in cardiac-vagal activity and peripheral sympathetic outflow.[52] A generally accepted thesis during the late 1970s and early 1980s was that arterial baroreceptor reflex regulation of blood pressure was diminished in the endurance-trained human. This was largely based upon the work of Stegemann et al.,[21] who reported on 50 young men in 1974. Twenty-five men were physically active long-distance runners, soccer players, and swimmers. Pressure was randomly varied from –60 to +60 mmHg in an air-tight chamber, which enclosed the subject's head and neck, and steady-state blood pressure changes (cuff method) and heart rate were determined about 3 min after each pressure change was completed. They noted that the reflex blood pressure changes produced during the extremes of the applied chamber pressure were less in the physically active compared with the sedentary group. They also noted that the reflex increases in the heart rate in response to decreases in carotid sinus transmural pressure (neck pressure) were less in the active compared with the sedentary individuals. Although this work is often cited, it has some deficiencies in design and controls. For example, no quantitative information on the level of aerobic fitness was obtained from these volunteers, nor was there a description of ages or baseline heart rate and blood pressures. The neck suction device employed by Stegemann and colleagues enclosed the entire head and neck region. Positive pressure applied to this device could have conceivably activated other reflexes, e.g., the oculocardiac reflex, which might have modified the assessment of carotid baroreflex-mediated responses. Furthermore, these investigators measured reflex responses 3 min after the onset of the head and neck stimuli. During these steady-state periods, carotid baroreflex responses should have been substantially modified by stimulation or adaptation of aortic and cardiopulmonary baroreflexes, and tracheal and somatic reflexes.

More detailed cross-sectional studies that lend support to Stegemann's work have been carried out by Smith and colleagues,[32] who studied 30 young male volunteers. Ten subjects were studied who had been involved in running competition for greater than 3 years and averaged greater than 50 mi/week in their training routine. They also studied ten subjects who competed in state or national weight-lifting competition and ten sedentary control individuals. Graded progressive infusions of phenylephrine were used to produce gradual increases in systemic blood pressure, and linear regression analyses were employed to determine the relationship between blood pressure elevations and heart rate reductions. They noted that baroreflex slopes of the sedentary group and the weight lifters were similar, but the slopes derived from the endurance-trained individuals were significantly (but subtly) decreased. One limitation of this study design was the use of phenylephrine to raise blood pressure. This limits the authors' conclusions to simply that endurance-trained athletes do not buffer hypertensive stimuli as well as nonathletes or weight lifters. These conclusions cannot be extended to orthostatic stress, which reduces blood pressure, since this limb of the baroreflex was not studied.

However, acute hypotensive stimuli have been delivered to the carotid sinus baroreceptors with a neck collar in several studies that have compared aerobically fit to sedentary individuals. Collectively, these studies indicate that carotid baroreceptor regulation of the heart rate is not impaired in the aerobically fit population. The influence of endurance training on carotid baroreflex responsiveness was first examined in a preliminary report by Falsetti et al.,[53] who used a neck collar to specifically activate or deactivate carotid sinus baroreceptors in nine competitively trained male swimmers (Vo_{2max} = 56 ml/kg/min) and nine age-matched untrained controls (Vo_{2max} = 42 ml/kg/min). Heart rate increases and decreases in response to neck pressure and suction were not significantly different between groups. A more recent neck collar study by Fiocchi et al.[54] examined 24 cycling tourists whose ages ranged from 16 to 60 years. They were unable to derive a significant relationship between baroreflex sensitivity and peak oxygen uptake. Unfortunately age, which is known to have a significant influence on baroreflex function, was not controlled for and the bicyclist's responses were not compared with a sedentary control group. However, in a later publication, Fiocchi et al.[55] reported that baroreflex sensitivity (determined with a neck collar) was greater in 11 distance runners (l0.3 ± 5.9 ms/mmHg) compared with l2 age-matched normal subjects (6.79 ± 3.8 ms/mmHg).

Research from our laboratory examined 20 healthy male subjects, l9 to 3l years of age.[43] These volunteers were grouped into a high-fit (Vo_{2max} > 53 ml/kg/min) or sedentary group (Vo_{2max} between 34 and 42 ml/kg/min). Baroreflex control of heart rate was determined with a neck collar in which the timing of the neck stimulus in the cardiac cycle, the rate of change (dP/dt) of the neck suction and pressure, and the duration of the carotid sinus stimulus were carefully controlled. Multiple repetitions of identical carotid stimuli were applied so that random, non-baroreceptor-mediated heart rate fluctuations could be diminished through averaging. R-R interval responses were assessed during the first several seconds after the onset of the carotid stimulus. Heart rate responses were determined in the first few seconds after the onset of neck suction or pressure, and were probably not substantially altered by opposing input from aortic baroreceptors, since blood pressure changes begin to occur 2 to 3 s after the onset of the carotid sinus stimulus. In this study, the reflex bradycardia of the high-fit individuals was significantly greater than that recorded from the unfit subjects. The reflex tachycardia in response to neck pressure was not different between the groups. Vo_{2max} was significantly correlated to baroreflex responsiveness. Although the mechanism of the enhanced baroreflex response to neck suction among the high-fit individuals was not specifically determined as part of the study, a significant direct relationship was derived between baroreflex-mediated bradycardia and respiratory sinus arrhythmia. Since respiratory sinus arrhythmia is known to be an index of baseline cardiac-vagal activity[56,57] and baroreflex slowing of heart rate in humans is primarily, if not solely, due to cardiac-vagal activity,[58,59] we believe our research findings support the possibility that cardiac-vagal activity was augmented in the high-fit group and that cardiac-vagal responsiveness to baroreceptor input was not attenuated, and was perhaps augmented, in the group of high-fit individuals chosen for study.

A recent report has been published by a separate lab that employed neck-collar techniques that were virtually identical to those employed in our research.[60] These authors also were unable to demonstrate a reduction in baroreflex responsiveness in endurance-trained athletes when compared with sedentary controls.

Recent longitudinal studies in both animals and humans support the notion that baroreflex responsiveness is not diminished (and may be augmented) in endurance-trained individuals. Billman et al.[61] defined a group of dogs with hearts that were susceptible to ventricular fibrillation. These dogs initially had low baroreflex sensitivities (phenylephrine method). However, aerobic training of these animals resulted in an improved gain of the baroreflex control of heart rate. A study by Somers et al.[62] assessed the baroreflex control of heart rate with bolus injections of phenylephrine in l6 borderline hypertensive patients. Eight of these patients were chosen to undergo a 6-month aerobic-training program. At the completion of this program, respiratory sinus arrythmia was increased and baroreflex sensitivity was augmented compared with the eight nontrained controls.

In summary, it is clear that the initial suggestion by Stegemann et al.[21] that baroreflex control of blood pressure is diminished in endurance-trained athletes is supported by only one study, which employed gradual infusions of phenylephrine to raise blood pressure. In contrast, research from separate laboratories, in which carotid sinus hypertension has been invoked by brief applications of neck suction or phenylephrine, indicates that reflex cardiac slowing is unchanged[53,54,60,61] or is augmented[43,55,62] in endurance-trained individuals compared with sedentary controls. When carotid sinus hypotension has been produced with neck pressure, baroreflex-mediated cardioacceleration has been shown to be similar between endurance-trained and sedentary individuals.[43,53,54]

There are no published studies that have compared reflex responses of endurance-trained and sedentary individuals to gradual reductions in systemic blood pressure produced by controlled infusions of vasodilators (e.g., sodium nitroprusside or nitroglycerine). This research approach would seem most appropriate, as vasodilators would create an environment similar to orthostatic stress, i.e., a gradual hypotension, which must be opposed by neural reflex and/or hormone responses. Several investigators have attempted to examine this portion of the baroreceptor reflex by determining the reduction in systolic pressure invoked by gradual applications of LBNP and relating this to the concomitant increase in heart rate. In several reports,[27,28] this index (change in systolic pressure/change in heart rate) during LBNP was less in trained compared with untrained individuals; however, in several separate studies, this index was either the same or greater in exercise-trained individuals when compared with sedentary control subjects.[10,63] Furthermore, this index was not significantly altered following a 12-week cycle ergometry conditioning program.[63]

D. ARTERIAL BAROREFLEX CONTROL OF PERIPHERAL RESISTANCE

Little information exists concerning reflex peripheral sympathetic responses to baroreceptor stimuli in humans. This is probably of equal importance to reflex heart rate responses when considering the regulation of blood pressure in humans. One technique to study these responses involves the application of 40 mmHg of LBNP, which simultaneously reduces cardiac filling pressure and arterial pressure, and provokes a reflex increase in forearm vascular resistance (and heart rate). When this stress was applied to endurance-trained swimmers and nonswimmers, no differences in the reflex vasoconstrictor responses of the forearm were noted.[64] However, when endurance-trained football players were studied, reflex increases in forearm vascular resistance were reported to be greater than those recorded from sedentary controls.[65] When LBNP of 40 mmHg was applied to individuals before and after a 12-week cycle exercise program, baroreflex-mediated forearm vascular resistance increases were reported to have diminished.[66] Moreover, Raven et al.[27] reported that peripheral vascular resistance increases provoked by increasing LBNP from 40 to 50 mmHg were less in high-fit compared with average-fit individuals. Because of these inconsistent findings on the effect of endurance training on forearm and peripheral vascular resistance responses to the simultaneous unloading of low- and high-pressure barore-ceptors with high levels of LBNP, one cannot draw any strong conclusions regarding the effects of high aerobic fitness on reflex regulation of peripheral vascular resistance.

E. LOW-PRESSURE (CARDIOPULMONARY) BARORECEPTOR REFLEXES

Low-pressure baroreceptors are located at the junction of the vena cava and right atrium, in the right and left atria, in the pulmonary veins, and within the walls of the heart. These are stretch receptors that respond to small changes of intrathoracic blood volume and initiate reflex alterations in splanchnic and muscle vascular resistance to maintain constant arterial blood pressure.[67] The function of these receptors in the endurance-trained athlete is not clearly known, however, there is reason to believe that the neural afferent traffic from these receptors might be altered with endurance training. For example, high levels of aerobic fitness are associated with elevations in plasma and blood volume.[14,33-35] This is thought to lead to chronic overstimulation of the low-pressure baroreceptors, since research findings indicate that baseline PRA[68,69] and plasma norepinephrine levels[70,71] are often lower in trained athletes compared with sedentary controls. In addition, ventricular mass is increased in endurance-trained athletes,[49] and this may alter the firing pattern of baroreceptors located within the ventricle. Early work by Boening in 1972[39] indicated that trained athletes had a delayed diuresis and a reduced natriuresis during water immersion (to the neck) compared with untrained controls. It was conjectured that this delay was due to a decreased sensitivity of the atrial volume baroreceptors, which are thought to be involved in antidiuretic hormone secretion in humans.

More specific information focusing on the reflex responses to unloading of low-pressure baroreceptors has been provided in studies employing low levels of LBNP (5 to 20 mmHg).

These levels of LBNP reduce cardiac filling pressures without lowering systemic blood pressure, which suggests that this stress may unload low-pressure baroreceptors with little alteration of arterial baroreceptor afferent traffic. Takeshita et al.[65] studied 14 young varsity football players and 16 nonathletes and compared their responses to 10 mmHg of LBNP. Reductions of central venous pressure were similar between groups, but reflex forearm vasoconstriction was greater in the athletes. A preliminary, longitudinal study performed by the same research group indicated that reflex forearm vascular resistance responses to 10 mmHg of LBNP in seven subjects were increased after a 4-month daily exercise program.[66] In contrast, a more recent study by Mack and colleagues[72] in five average-fit subjects (Vo_{2max} = 38.5 ml/kg/min) and five high-fit subjects (Vo_{2max} = 57 ml/kg/min) demonstrated that when central venous pressure was reduced by graded low-level LBNP, the calculated reflex increases in forearm vascular resistance were less in the high-fit compared with the average-fit group. These results were the opposite of those found by Takeshita et al.,[65] and the authors suggest that this may have been due to differences in study populations (endurance vs. weight-trained subjects). An interesting preliminary observation was advanced by Falsetti et al.,[64] who reported that forearm vascular resistance increases provoked separately by either neck pressure or LBNP were not different between swimmers and sedentary individuals. However, when neck pressure and LBNP were applied simultaneously, the reflex increases of forearm vascular resistance in the untrained individuals were augmented more than those of the trained subjects. These findings suggest there may be some inhibitory or antagonistic effect of the central processing of afferent neural input during the simultaneous unloading of low- and high-pressure baroreceptors in the trained athlete.

F. VASOACTIVE HORMONE RESPONSES TO ORTHOSTATIC STRESS

Although acute cardiovascular adjustments to orthostatic stress are mediated by baroreceptors, the ability to withstand prolonged exposure to $+G_z$ forces, head-up tilt, or LBNP may depend upon adequate secretion of vasoactive hormones. Several vasoconstrictor hormones of importance include AVP, angiotensin II (part of the renin, angiotensin-aldosterone axis), and norepinephrine. Studies in healthy volunteers have demonstrated that head-up tilt, LBNP, or thigh cuff inflation elicit immediate significant increases in plasma norepinephrine levels,[73,74] but these stresses must be maintained for longer periods of time to evoke detectable changes in plasma renin and angiotensin levels.[74-76] Plasma vasopressin remains relatively unchanged during orthostatic stress until presyncopal signs (sudden bradycardia or large reductions in blood pressure) occur.[73,77]

Resting supine plasma norepinephrine[70,71] and renin[68,69] levels have been reported to be lower in endurance-trained subjects compared with untrained controls. This may partly be due to the reported increased blood volume[14,33-35] and probable heightened stimulation of low-pressure baroreceptors in the trained athlete. An interesting preliminary report by Sather et al.[78] suggests that subjects with a high tolerance to LBNP have significantly greater AVP and PRA levels at syncope compared with low-tolerance subjects. The endurance capacity of these subjects was not reported. However, a preliminary report by Goldwater et al.[79] suggests that athletically conditioned middle-age individuals with low orthostatic tolerance not only had low baseline levels of PRA, but also demonstrated reduced increases in PRA in response to LBNP compared with those of sedentary controls. Moreover, Convertino et al.[7] provides graphic data suggesting that plasma norepinephrine level increases produced by LBNP are less in high-aerobic fit versus low-aerobic fit subjects. Collectively, these reports suggest that the secretion of AVP, renin, and norepinephrine may be diminished during orthostasis in high-fit individuals. Since these vasoactive hormones are under baroreceptor control, potentially interesting research would be to determine the secretion of these hormones during controlled, *quantifiable* unloading of baroreceptors in aerobically fit and unfit individuals. To our knowledge, this has not been done.

V. SUMMARY

The question of whether aerobic fitness contributes to orthostatic intolerance is clearly not resolved. This may be partially due to the fact that there are important limitations in many of the studies that have attempted to relate aerobic fitness and orthostatic intolerance. For example, the reduction in central blood volume produced by orthostatic stress has not been quantified in any study that has compared the aerobically fit to the sedentary individual. Many studies have measured calf circumference increases or calf or thigh impedance decreases during orthostatic stress, but these methods provide only limited information and may not precisely reflect central events. They do not permit quantification of blood pooling in the splanchnic or pelvic regions, which serve as important blood reservoirs in the human body. At the very least, a reasonable indicator of central hemodynamics during orthostatic stress would be the measurement of central venous pressure. Another less invasive technique would be to use echocardiography to assess atrial and ventricular volumes during LBNP or tilt. With these methods, cardiovascular responses could be plotted as a function of a true independent variable instead of simply the level of LBNP or the angle of tilt.

Another limitation of the majority of previously quoted studies has been the lack of control, or determination, of salt and water balance prior to initiating studies. It is conceivable that the athlete who comes to the lab shortly after completing a 10-mi run or a 1-h cycling session may respond differently to orthostatic tests than a similar trained athlete whose last training session was the day prior to experimentation and who has had an overnight period to replenish fluid and electrolytes.

The majority of the cross-sectional studies and longitudinal studies that have refuted an association between aerobic fitness and orthostatic tolerance have been deficient in their lack of control for the type of exercise employed to achieve aerobic fitness. A few studies that have found inadequate regulation of blood pressure during orthostatic stress in the aerobically fit have primarily been those that focused on endurance-trained *runners*. Several factors appear to predispose runners to syncope. First, they have a small ratio of muscle to cross-sectional area of the calf.[38] This promotes a greater degree of caudal venous blood pooling during orthostatic stress.[10,18,23,28,29] In contrast, endurance-trained cyclists, skaters, skiers, and swimmers have a larger muscle mass of the lower extremities, which mechanically restricts venous distension and blood pooling. Secondly, runners appear to have attenuated peripheral vasoconstriction during LBNP[27] or tilt.[26] This promotes hypotension and probably contributes to reduced venous constriction. This deficient vasoconstrictor response may be partly related to a blunted increase in circulating levels of norepinephrine,[7] renin,[78] and perhaps vasopressin[78] during orthostatic stress. The altered regulation of peripheral resistance and secretion of vasoactive hormones during orthostasis may be related to training-induced changes in the function of the low-pressure cardiopulmonary baroreceptors. It has been postulated that the increase in circulating blood volume that occurs with endurance training could, over time, lead to changes in receptor function. It is also conceivable that the increase in ventricular mass of the endurance-trained athlete may alter the firing characteristics of these receptors.

One factor that has limited our ability to resolve the question as to whether aerobic fitness contributes to orthostatic intolerance has been the fact that we still are uncertain as to what is the trigger for syncope. There is a large list of iatrogenic causes and pathological conditions that lead to orthostatic hypotension, but in the healthy young volunteer who is free of systemic diseases, what causes syncope? Is it simply cerebral hypoperfusion, or is it due to activation of vagal afferent fibers from the heart[80] due to strong contractions of an empty ventricle? Either of these mechanisms would come into play if venous return was profoundly reduced during orthostatic stress. The endurance-trained runner is predisposed to orthostatic intolerance due to multiple factors, which promote reduced venous return, a decrease in cardiac output, and a reduction in blood pressure (Table 1).

TABLE 1
Probable Contributors to Orthostatic Intolerance in the Aerobically Fit Runner

Altered low-pressure baroreceptor function secondary to:
   Increased ventricular mass
   Increased blood volume
Diminished peripheral vasoconstriction
Decreased relative peak heart rate increases
Reduced release of vasoactive hormones
Reduced lower extremity muscle mass surrounding veins

Future studies should focus on the elite endurance-trained runner. Longitudinal studies may be inadequate in that the physiological changes that occur in the orthostatic intolerant athlete may require many years of training to produce anatomical and physiological adaptations that contribute to orthostatic hypotension. Future studies must give consideration to the fluid and electrolyte status of each research volunteer. The precise degree of orthostatic stress (reduction in central blood volume) must be quantified in some invasive or noninvasive fashion. Beat-to-beat cardiovascular responses and frequent blood sampling for vasoactive hormones should be included as part of each testing protocol. Additional groups of endurance-trained athletes who are not runners should be included separately as part of these studies. We believe that such carefully controlled studies that employ quantifiable methods of monitoring will improve our understanding of how aerobic fitness influences the maintenance of blood pressure and consciousness during orthostatic stress in humans.

REFERENCES

1. **Crampton, W. C.,** The gravity resisting ability of the circulation. Its measurement and significance (blood ptosis), *Am. J. Med. Sci.*, 160, 721, 1920.
2. **Turner, A. H.,** The circulatory minute volume of healthy young women in reclining, sitting and standing positions, *Am. J. Physiol.*, 80, 605, 1927.
3. **Shvartz, E.,** Relationship between endurance and orthostatic tolerance, *J. Sports Med.*, 8, 75, 1968.
4. Shvartz, E., Meroz, A., Magazanik, A., Shoenfeld, Y., and Shapiro, Y., Exercise and heat orthostatism and the effect of heat acclimation and physical fitness, Aviat. Space Environ. Med., 48, 836, 1977.
5. **Allen, S. C., Taylor, C. L., and Hall V. E.,** A study of orthostatic inefficiency by tilt board method, *Am. J. Physiol.*, 143, 11, 1945.
6. **Mayerson, H. S.,** Roentgenkymographic determination of cardiac output in syncope induced by gravity, *Am. J. Physiol.*, 138, 630, 1942.
7. **Convertino, V. A.,** Aerobic fitness, endurance training, and orthostatic intolerance, *Exercise Sport Sci. Rev.*, 15, 223, 1987.
8. **Klein, K. E., Bruner, H., Jovy, D., Vogt, L., and Wegmann, H. M.,** Influence of stature and physical fitness on tilt-table and acceleration tolerance, *Aerospace Med.*, 40, 293, 1969.
9. **Klein, K. E., Wegmann, H. M., Bruner, H., and Vogt, L.,** Physical fitness and tolerances to environmental extremes, *Aerospace Med.*, 40, 998, 1969.
10. **Hudson, D. L., Smith, M. L., Graitzer, H., and Raven, P. B.,** Fitness-related differences in response to LBNP during sympathetic blockade with metoprolol, *Fed. Proc.*, 45 (Abstr.), 643, 1986.
11. **Sather, T. M., Goldwater, D. J., Montgomery, L. D., and Convertino, V. A.,** Cardiovascular dynamics associated with tolerance to lower body negative pressure, *Aviat. Space Environ. Med.*, 57, 413, 1986.
12. **Frey, M. A. B., Mathes, K. L., and Hoffler, G. W.,** Relationship between aerobic fitness and responses to lower-body negative pressure in women, *Aviat. Space Environ. Med.*, 58, 1149, 1987.
13. **Convertino, V. A., Sather, T. M., Goldwater, D. J., and Alford, W. R.,** Aerobic fitness does not contribute to prediction of orthostatic intolerance, *Med. Sci. Sports Exercise*, 18, 551, 1986.

14. **Kjellberg, S. R., Rudhe, U., and Sjostrand, T.,** Increase of the amount of hemoglobin and blood volume in connection with physical training, *Acta Physiol. Scand.*, 19, 146, 1950.

15. **Ludwig, D. A., Convertino, V. A., Goldwater, D. J., and Sandler, H.,** Logistic risk model for the unique effects of inherent aerobic capacity and $+G_z$ tolerance before and after simulated weightlessness, *Aviat. Space Environ. Med.*, 58, 1057, 1987.

16. **Cooper, K. H. and Leverett S., Jr.,** Physical conditioning versus $+G_z$ tolerance, *Aerospace Med.*, 37, 462, 1966.

17. **Sheldahl, L. M., Tristani, F. E., Barney, J. A., Groban, L., Levandoski, S., Christie, J., and Smith, J. J.,** Effect of endurance training on hemodynamic response to postural stress, *Fed. Proc.*, 45 (Abstr.), 282, 1986.

18. **Convertino, V. A., Montgomery, L. D., and Greenleaf, J. E.,** Cardiovascular responses during orthostasis: effect of an increase in VO_{2maxx}, *Aviat. Space Environ. Med.*, 55, 702, 1984.

19. **Greenleaf, J. E., Dunn, E. R., Nesvig, C., Keil, L. C., Harrison, M. H., Geelen, G., and Kravik, S. E.,** Effect of longitudinal physical training and water immersion on orthostatic tolerance in men, *Aviat. Space Environ. Med.*, 59, 152, 1988.

20. **Lightfoot, J. T., Clayton, R. P., Torok, D. J., Fournell, T. W., and Fortney, S. M.,** Ten weeks of aerobic training do not affect lower body negative pressure responses, *J. Appl. Physiol.*, 67, 894, 1989.

21. **Stegemann, J., Busert, A., and Brock, D.,** Influence of fitness on the blood pressure control system in man, *Aerospace Med.*, 45, 45, 1974.

22. **Greenleaf, J. E., Haines, R. F., Bernauer, E. M., Morse, J. T., Sandler, H., Armbruster, R., Sagan, L, and van Beaumont, W.,** $+G_z$ tolerance in man after 14-day bedrest periods with isometric and isotonic exercise conditioning, *Aviat. Space Environ. Med.*, 46, 671, 1975.

23. **Luft, U. C., Myhre, L. G., Loeppky, J. A., and Venters, M. D.,** A study of factors affecting tolerance of gravitational stress simulated by lower body negative pressure, in Research Report on Specialized Physiology Studies in Support of Manned Space Flight, NASA Contract NAS 9-14472, Lovelace Foundation, Albuquerque, NM, 1976, 1.

24. **Luft, U. C., Loeppky, J. A., Venters, M. D., Greene, E. R., Eldridge, M. W., Hoekenga, D. E., and Richards, K. L.,** Tolerance of lower body negative pressure (LBNP) in endurance runners, weightlifters, swimmers, and nonathletes, in Research Report on Specialized Physiology Studies in Support of Manned Space Flight, NASA Contract NAS 9-15483, Lovelace Foundation, Albuquerque, NM, 1980, 1.

25. **Klein, K. E., Wegmann, H. M., and Kuklinski, P.,** Athletic endurance training — advantage for spaceflight?: the significance of physical fitness for selection and training of Spacelab crews, *Aviat. Space Environ. Med.*, 48, 215, 1977.

26. **Mangseth, G. R. and Bernauer, E. M.,** Cardiovascular response to tilt in endurance trained subjects exhibiting syncopal reactions, *Med. Sci. Sports Exerc.*, 12 (Abstr.), 140, 1980.

27. **Raven, P. B., Rohm-Young, D., and Blomqvist, C.,** Physical fitness and cardiovascular response to lower body negative pressure, *J. Appl. Physiol.*, 56, 138, 1984.

28. **Smith, M. L. and Raven P. B.,** Cardiovascular responses to lower body negative pressure in endurance and static exercise-trained men, *Med. Sci. Sport Exerc.*, 18, 545, 1986.

29. **Pawelczyk, J. A., Kenney, W. L., and Kenney, P.,** Cardiovascular responses to head-up tilt after an endurance exercise program, *Aviat. Space Environ. Med.*, 59, 107, 1988.

30. **Epperson, W. L., Burton, R. R., and Bernauer, E. M.,** The influence of differential physical conditioning regimens on simulated aerial combat maneuvering tolerance, *Aviat. Space Environ. Med.*, 53, 1091, 1982.

31. **Convertino, V. A., Doerr, D. F., Flores, J. F., Hoffler, G. W., and Buchanan, P.,** Leg size and muscle functions associated with leg compliance, *J. Appl. Physiol.*, 64, 1017, 1988.

32. **Smith, M. L., Graitzer, H. M., Hudson, D. L., and Raven, P. B.,** Baroreflex function in endurance-and static exercise-trained men, *J. Appl. Physiol.*, 64, 585, 1988.

33. **Oscai, L. B., Williams, B. T., and Hertig, B. A.,** Effect of exercise on blood volume, *J. Appl. Physiol.*, 24, 622, 1968.

34. **Holmgren, A., Mossfeldt, F., Sjostrand, T., and Strom, G.,** Effect of training on work capacity, total hemoglobin, blood volume, heart volume and pulse rate in recumbent and upright positions, *Acta Physiol. Scand.*, 50, 72, 1960.

35. **Convertino, V. A., Brock, P. J., Keil, L. C., Bernauer, E. M., and Greenleaf, J. E.,** Exercise training-induced hypervolemia: role of plasma albumin, renin and vasopressin, *J. Appl. Physiol.*, 48, 665, 1980.

36. **Mifflin, S. W. and Kunze, D. L.,** Rapid resetting of low-pressure vagal receptors in the superior vena cava of the rat, *Circ. Res.*, 51, 241, 1982.

37. **Harrison, M. H., Keil, L. C., Wade, C. A., Silver, J. E., Geelen, G., and Greenleaf, J. E.,** Effect of hydration on plasma volume and endocrine responses to water immersion, *J. Appl. Physiol.*, 61, 1410, 1986.

38. **Norsk, P. and Epstein, M.,** Effects of water immersion on arginine vasopressin release in humans, *J. Appl. Physiol.*, 64, 1, 1988.

39. **Boening, D., Ulmer, H-V., Meier, U., Skipka, W., and Stegemann, J.,** Effects of multi-hour immersion on trained and untrained subjects: I. Renal function and plasma volume, *Aerospace Med.*, 43, 300, 1972.

40. **Clausen, J. P.,** Effect of physical training on cardiovascular adjustments to exercise in man, *Physiol. Rev.*, 57, 779, 1977.

41. **Ekblom, B., Kilbom, A., and Soltysiak, J.,** Physical training, bradycardia, and autonomic nervous system, *Scand. J. Clin. Lab. Invest.*, 32, 251, 1973.

42. **Frick, M. H., Elovainio, R. O., and Somer, T.,** The mechanism of bradycardia evoked by physical training, *Cardiologia*, 51, 46, 1967.

43. **Barney, J. A., Ebert, T. J., Groban, L., Farrell, P. A., Hughes, C. V., and Smith, J. J.,** Carotid baroreflex responsiveness in high-fit and sedentary young men, *J. Appl. Physiol.*, 65, 2190, 1988.

44. **Ostman-Smith, I.,** Adaptive changes in the sympathetic venous system and some effector organs of the rat following long-term exercise or cold acclimation and the role of cardiac sympathetic nerves in the genesis of compensatory cardiac hypertrophy, *Acta Physiol. Scand.*, 477 (Suppl.), 1, 1979.

45. **Davell, R. T. and Tipton, C. M.,** Influence of training on the heart rate responses of rats to isoproterenol and propranolol, *Physiologist*, 13 (Abstr.), 182, 1970.

46. **Tipton, C. M. and Taylor, B.,** Influence of atropine on heart rate of rats, *Am. J. Physiol.*, 208, 480, 1965.

47. **Lewis, S. F., Nylander, E., Gad, P., and Areskog, N. H.,** Non-autonomic component in bradycardia of endurance-trained men at rest and during exercise, *Acta Physiol. Scand.*, 109, 297, 1980.

48. **Hammond, H. K., White, F. C., Brunton, L. L., and Longhurst, J. C.,** Association of decreased myocardial B-receptors and chronotropic response to isoproterenol and exercise in pigs following chronic dynamic exercise, *Circ. Res.*, 60, 720, 1987.

49. **Scheuer, J. and Tipton, C. M.,** Cardiovascular adaptations to physical training, *Ann. Rev. Physiol.*, 39, 221, 1977.

50. **Svedenhag, J.,** The sympatho-adrenal system in physical conditioning. Significance for training-induced adaptations and dependency on the training state, *Acta Physiol. Scand.*, 125 (Suppl.), 1, 1985.

51. **Raven, P. B. and Smith, M. L.,** Physical fitness and its effect on factors affecting orthostatic tolerance, *The Physiologist*, 27 (Suppl.), S59, 1984.

52. **Mancia, G. and Mark, A. L.,** Arterial baroreflexes in humans, in *Handbook of Physiology, Sect. 2: The Cardiovascular System, Vol. III: Peripheral Circulation and Organ Blood Flow, Part 2*, Shepherd, J. T. and Abboud, F. M. Eds., American Physiological Society, Bethesda, MD, 1983, 755.

53. **Falsetti, H., Burke, E., and Tracy, J.,** Cardiopulmonary and carotid baroreflex control of blood pressure in athletes, *Clin. Res.*, 30 (Abstr.), A759, 1982.

54. **Fiocchi, R., Fagard, R., Vanhees, L., Grauwels, R., and Amery, A.,** Carotid baroreflex sensitivity and physical fitness in cycling tourists, *Eur. J. Appl. Physiol.*, 54, 461, 1985.

55. **Fiocchi, R., Fagard, R., Staessen, J., Vanhees, L., and Amery, A.,** Atrioventricular block induced in an athlete by carotid baroreceptor stimulation, *Am. Heart J.*, 109, 1102, 1985.

56. **Eckberg, D. L.,** Human sinus arrhythmia as an index of vagal cardiac outflow, *J. Appl. Physiol.*, 54, 961, 1983.

57. **Katona, F. G. and Jih, F.,** Respiratory sinus arrhythmia: non-invasive measure of parasympathetic cardiac control, *J. Appl. Physiol.*, 39, 801, 1975.

58. **Eckberg, D. L.,** Nonlinearities of the human carotid baroreceptor-cardiac reflex, *Circ. Res.*, 47, 208, 1980.

59. **Leon, D. F., Shaver, J. A., and Leonard, J. J.,** Reflex heart rate control in man, *Am. Heart J.*, 80, 729, 1970.

60. **Reiling, M. J. and Seals, D. R.,** Respiratory sinus arrhythmia and carotid baroreflex control of heart rate in endurance athletes and untrained controls, *Clin. Physiol.*, 8, 511, 1988.

61. **Billman, G. E., Schwartz, P. J., and Stone, H. L.,** The effects of daily exercise on susceptibility to sudden cardiac death, *Circulation*, 69, 1182, 1984.

62. **Somers, V. K., Conway, J., and Sleight, P.,** The effect of physical training on blood pressure and baroreflex sensitivity, *Circulation*, 76 (Abstr.), IV-61, 1987.

63. **Vroman, N. B., Healy, J. A., and Kertzer, R.,** Cardiovascular responses to lower body negative pressure (LBNP) following endurance training, *Aviat. Space Environ. Med.*, 59, 330, 1988.

64. **Falsetti, H. L., Burke, E. R., and Tracy, J.,** Carotid and cardiopulmonary baroreflex control of forearm vascular resistance in swimmers, *Med. Sci. Sports Exerc.*, 15 (Abstr.), 183, 1983.

65. **Takeshita, A., Jingu, S., Imaizumi, T., Kunihiko, Y., Koyanagi, S., and Nakamura, M.,** Augmented cardiopulmonary baroreflex control of forearm vascular resistance in young athletes, *Circ. Res.*, 59, 43, 1986.

66. **Jingu, S., Takeshita, A., Ashihara, T., Imaizumi, T., Yamamota, K., Hoka, S., Ito, N., and Nakamura, M.,** Exercise augments cardiopulmonary baroreflex in man, *Circulation*, 68 (Suppl. 3) (Abstr.), 77, 1983.

67. **Mark, A. L. and Mancia, G.,** Cardiopulmonary baroreflexes in humans, in *Handbook of Physiology, Sect. 2, Vol. III, Part 2*, Shepherd, J. T. and Abboud, F. M. Eds., American Physiological Society, Bethesda, MD, 1983, 795.

68. **Geyssant, A., Geelen, G., Denis, C., Allevard, A. M., Vincent, M., Jarsaillon, E., Bizollon, C. A., Lacour, J. R., and Gharib, C.,** Plasma vasopressin, renin activity, and aldosterone: Effect of exercise and training, *Eur. J. Appl. Physiol.*, 46, 21, 1981.

69. **M'Buyamba-Kabangu, J. R., Fagard, R., Lijnen, P., and Amery, A.,** Relationship between plasma renin activity and physical fitness in normal subjects, *Eur. J. Appl. Physiol.*, 53, 304, 1985.

70. **Hartley, L. H., Mason, J. W., Hogan, R. P., Jones, L. G., Kotchen, T. A., Mougey, E. H., Wherry, F. E., Pennington, L. L., and Ricketts, P. T.,** Multiple hormonal responses to graded exercise in relation to physical training, *J. Appl. Physiol.*, 33, 602, 1972.

71. **Kiyonaga, A., Arakawa, K., Tanaka, H., and Shindo, M.,** Blood pressure and hormonal responses to aerobic exercise, *Hypertension*, 7, 125, 1985.

72. **Mack, G. W., Shi, X., Nose, H., Tripathi, A., and Nadel, E. R.,** Diminished baroreflex control of forearm vascular resistance in physically fit humans, *J. Appl. Physiol.*, 63, 105, 1987.

73. **Goldsmith, S. R., Francis, G. S., Cowley, A. W., and Cohn, J. N.,** Response of vasopressin and norepinephrine to lower body negative pressure in humans, *Am. J. Physiol.*, 243, H970, 1982.

74. **Egan, B., Fitzpatrick, M. A., and Julius, S.,** The heart and regulation of renin, *Circulation*, 75 (Suppl. I), I-130, 1987.

75. **Mark, A. L., Abboud, F. M., and Fitz, A. E.,** Influence of low- and high-pressure baroreceptors on plasma renin activity in humans, *Am. J. Physiol.*, 235, H29, 1978.

76. **Oparil, S., Vassaux, C., Sanders, C. A., and Haber, E.,** Role of renin in acute postural homeostasis, *Circulation*, 41, 89, 1970.

77. **Davies, R., Slater, J. D. H., Forsling, M. L., and Payne, N.,** The response of arginine vasopressin and plasma renin to postural change in normal man, with observations on syncope, *Clin. Sci. Mol. Med.*, 51, 267, 1976.

78. **Sather, T. M., Convertino, V. A., Goldwater, D. J., Keil, L. C., Kates, R., and Montgomery, L. D.,** Vasoactive neuroendocrine responses associated with orthostatic tolerance in man, *Fed. Proc.*, 44 (Abstr.), 817, 1985.

79. **Goldwater, D. J., DeLada, M., Polese, A., Keil, L., and Leutscher, J. A.,** Effect of athletic conditioning on orthostatic tolerance after prolonged bedrest, *Circulation*, 62 (Abstr.), III-287, 1980.

80. **Oberg, B. and White, S.,** The role of vagal cardiac nerves and arterial baroreceptors in the circulatory adjustments to hemorrhage in the cat. *Acta Physiol. Scand.*, 80, 395, 1970.

Chapter 3

WEIGHTLESSNESS AND RESPONSE TO ORTHOSTATIC STRESS

Mary Anne Bassett Frey, John B. Charles, and Diana E. Houston

TABLE OF CONTENTS

I. INTRODUCTION

In the nearly weightless environment of space flight, hydrostatic pressure gradients do not exist, and the distribution of body fluids is independent of body position or posture. Thus, reflex cardiovascular responses are not needed to maintain arterial blood pressure during postural changes. Why, then, should a chapter on weightlessness be included in this book about circulatory response to the upright posture? Exposure to weightlessness alters the cardiovascular system's response to provocative orthostatic-like stress during weightless flight and diminishes the ability to respond to changes in posture upon return to Earth's gravitational environment (1 G). Furthermore, although crewmembers have withstood the stress of reentry to date, they have consistently experienced altered orthostatic function after even brief missions. This chapter is being included because knowledge of the processes of this adaptation to weightlessness and readaptation to gravitational forces should enhance our understanding of cardiovascular regulatory mechanisms on Earth. Furthermore, the study of human cardiovascular function in actual and simulated weightlessness has implications for the future of both space travel and the practice of clinical medicine.

The two major purposes of this chapter are (1) to summarize what is known about cardiovascular and body fluid changes in weightlessness that might affect orthostatic function and (2) to describe the evidence of altered orthostatic function during and after space flight. Data on physiological responses during flight are limited. Thus, we have relied heavily on ground-based simulations of weightlessness to suggest the mechanisms by which orthostatic function changes during space flight. The value of simulation studies depends on how well the test conditions evoke the same physiological changes that occur during actual weightlessness, particularly the cephalad redistribution of body fluids.

To facilitate the reader's understanding of the data from space flight, we present first a brief overview of human space flights through May 1989. These flights are listed in Tables 1 through 4.

II. OVERVIEW OF U.S. AND SOVIET HUMAN SPACE FLIGHTS THROUGH MAY 1989

The evidence presented in this chapter suggests that cardiovascular adjustments to space flight occur rapidly at first, then more gradually, such that after some finite time in flight, subsequent changes result more from in-flight activities than from exposure to microgravity. Investigators in the U.S.S.R. have placed the dividing point at about 1 month into the flight (John B. Charles, personal communication). For this reason, we distinguish between flights of 28 d or longer ("long-duration") and those of fewer days ("short-duration") in our tabulation of flights.

In the U.S. space program, short-duration missions (Table 1) have lasted from 15 min to 13.8 d. They include 6 Mercury missions (1962-1963), 10 Gemini missions (1965-1966), 12 Apollo missions (1968-1975, including the 1975 Apollo-Soyuz Test Project [ASTP]), and 28 missions of the present National Space Transportation System (NSTS) from 1981 through May 1989. The Apollo Program exposed 12 astronauts to the moon's gravitational forces of approximately one-sixth those on Earth as well as to weightless space flight.

TABLE 1
U.S. Manned Flights Lasting Fewer Than 28 Days[1,2]

Project	Mission	Launch year	Flight duration	Crew
Mercury	MR-3	1961	15 min (suborbital)	Shepard
	MR-4	1961	16 min (suborbital)	Grissom
	MA-6	1962	4.9 h	Glenn
	MA-7	1962	4.9 h	Carpenter
	MA-8	1962	9.2 h	Schirra
	MA-9	1963	1.4 d	Cooper
Gemini	GT-3	1965	4.9 h	Grissom, Young
	GT-4	1965	4.1 d	McDivitt, White
	GT-5	1965	8.0 d	Cooper, Conrad
	GT-7	1965	13.8 d	Borman, Lovell
	GT-6	1965	1.1 d	Schirra, Stafford
	GT-8	1966	10.7 h	Armstrong, Scott
	GT-9	1966	3.0 d	Stafford, Cernan
	GT-10	1966	2.9 d	Young, Collins
	GT-11	1966	3.0 d	Conrad, Gordon
	GT-12	1966	3.9 d	Lovell, Aldrin
Apollo	7	1968	10.8 d	Schirra, Eisele, Cunningham
	8	1968	6.1 d	Borman, Lovell, Anders
	9	1969	10.0 d	McDivitt, Scott, Schweickart
	10	1969	8.0 d	Stafford, Young, Cernan
	11	1969	8.1 d	Armstrong, Collins, Aldrin
	12	1969	10.2 d	Conrad, Gordon, Bean
	13	1970	6.0 d	Lovell, Swigert, Haise
	14	1971	9.0 d	Shepard, Roosa, Mitchell
	15	1971	12.3 d	Scott, Worden, Irwin
	16	1972	10.5 d	Mattingly, Young, Duke
	17	1972	12.6 d	Cernan, Evans, Schmitt
Apollo-Soyuz Test Project	18	1975	9.1 d	Stafford, Brand, Slayton
STS	1	1981	2.3 d	Young, Crippen
	2	1981	2.3 d	Truly, Engle
	3	1982	8.0 d	Lousma, Fullerton
	4	1982	7.0 d	Mattingly, Hartsfield
	5	1982	5.1 d	Brand, Overmyer, Lenoir, Allen
	6	1983	5.0 d	Weitz, Bobko, Musgrave, Peterson
	7	1983	6.1 d	Crippen, Hauck, Ride, Fabian, Thagard
	8	1983	6.0 d	Truly, Brandenstein, Bluford, D. Gardner, W. Thorton
	9	1983	10.3 d	Young, Shaw, Garriott, Parker, Lichtenberg, Merbold
	41B	1984	8.0 d	Brand, Gibson, McNair, McCandless, Stewart
	41C	1984	7.0 d	Crippen, Scobee, Hart, Nelson, Van Hoften
	41D	1984	6.0 d	Hartsfield, Coats, Hawley, Mullane, Resnik, C. Walker
	41G	1984	8.2 d	Crippen, McBride, Leestma, Sullivan, Ride, Scully-Power, Garneau
	51A	1984	8.0 d	Hauck, D. Walker, A. Fisher, Allen, D. Gardner

TABLE 1 (continued)
U.S. Manned Flights Lasting Fewer Than 28 Days[1,2]

Project	Mission	Launch year	Flight duration	Crew
	51C	1985	3.1 d	Mattingly, Shriver, Onizuka, Buchli, Payton
	51D	1985	7.0 d	Bobko, Williams, Seddon, Griggs, Hoffman, Garn, C. Walker
	51B	1985	7.0 d	Overmyer, Gregory, Lind, Thornton, Wang, Thagard, van den Berg
	51G	1985	7.1 d	Brandenstein, Creighton, Fabian, Lucid, Nagel, Baudry, Al-Saud
	51F	1985	7.9 d	Fullerton, Bridges, Henize, Musgrave, England, Acton, Bartoe
	51I	1985	7.1 d	Engle, Covey, Van Hoften, W. Fisher, Lounge
	51J	1985	4.1 d	Bobko, Grabe, Hilmers, Stewart, Pailes
	61A	1985	7.0 d	Hartsfield, Nagel, Dunbar, Bluford, Messerschmid, Furrer, Ockels, Buchli
	61B	1985	6.9 d	Shaw, O'Connor, Vela, C. Walker, Cleave, Ross, Spring
	61C	1986	6.1 d	Gibson, Bolden, Hawley, G. Nelson, Cenker, Chang-Diaz, B. Nelson
	26	1988	4.0 d	Hauck, Covey, Hilmers, Lounge, G. Nelson
	27	1988	4.4 d	R. Gibson, G. Gardner, Mullane, Ross, Shepard

TABLE 2
Soviet Manned Orbital Flights Lasting Fewer Than 28 Days[1]

Project	Mission	Launch year	Flight duration[a]	Crew
Vostok	1	1961	1.8 h	Gagarin
	2	1961	1.1 d	Titov
	3	1962	3.9 d	Nikolayev
	4	1962	3.0 d	Popovich
	5	1963	5.0 d	Bykovsky
	6	1963	3.0 d	Tereshkova
Voskhod	1	1964	1.0 d	Komarov, Feoktistov, Yegorov
	2	1965	1.1 d	Belyayev, Leonov
Soyuz	1	1967	1.1 d	Komarov (died)
	3	1968	4.0 d	Beregovoi
	4	1969	3.0 d	Shatalov
	5	1969	3.0 d	Volynov, Khrunov, Yeliseyev
	6	1969	4.9 d	Shonin, Kubasov
	7	1969	4.9 d	Filipchenko, Volkov, Gorbatko
	8	1969	4.9 d	Shatalov, Yeliseyev
	9	1970	17.7 d	Nikolayev, Sevastyanov
	12	1973	2.0 d	Lazarev, Makarov
	13	1973	7.9 d	Klimuk, Lebedev
	16	1974	5.9 d	Filipchenko, Rukavishnikov

TABLE 2 (continued)
Soviet Manned Orbital Flights Lasting Fewer Than 28 Days[1]

Project	Mission	Launch year	Flight duration[a]	Crew
	22	1976	7.9 d	Bykovsky, Aksyonov
Salyut 1	Soyuz 10	1971	2.0 d	Shatalov, Yeliseyev, Rukavishnikov, (docked 5.5 hours)
	Soyuz 11	1971	23.0 d	Dobrovolsky, Volkov, Patsayev (all died)
Salyut 3	Soyuz 14	1974	15.7 d	Popovich, Artyukhin
	Soyuz 15	1974	2.0 d	Sarafanov, Demin (failed to dock)
Apollo-Soyuz Test Project	Soyuz 19	1975	5.9 d	Kubasov, Leonov
Salyut 5	Soyuz 23	1976	2.0 d	Zudov, Rozhdestvensky (failed to dock)
	Soyuz 24	1977	17.7 d	Gorbatko, Glazkov
Salyut 6	Soyuz 25	1977	2.0 d	Kovalyonok, Ryumin (failed to dock)
	Soyuz 27	1978	6.0 d	Dzhanibekov, Makarov (visiting crew)
	Soyuz 28	1978	7.9 d	Gubarev, Remek (visiting crew)
	Soyuz 30	1978	7.9 d	Klimuk, Hermaszewski (visiting crew)
	Soyuz 31	1978	7.9 d	Bykovsky, Jaehn (visiting crew)
	Soyuz 33	1979	2.0 d	Rakavishnikov, Ivanov (failed to dock)
	Soyuz 36	1980	7.9 d	Kubasov, Farkas (visiting crew)
	Soyuz T-2	1980	3.9 d	Malyshev, Aksyonov (vehicle test)
	Soyuz 37	1980	7.9 d	Gorbatko, Tuan (visiting crew)
	Soyuz 38	1980	7.9 d	Romanenko, Tamayo-Mendez (visiting crew)
	Soyuz T-3	1980	12.8 d	Kizim, Makarov, Strekalov (maintenance mission)
	Soyuz 39	1981	7.9 d	Dzhanibekov, Gurragcha (visiting crew)
	Soyuz 40	1981	7.9 d	Popov, Prunariu (last visiting crew)
Salut 7	Soyuz T-6	1982	7.9 d	Dzanibekov, Ivenchenkov, Chretien (visiting crew)
	Soyuz T-7	1982	7.9 d	Savitskaya, Popov, Serebrov (visiting crew)
	Soyuz T-8	1983	2.0 d	Titov, Strekalov, Serebrov (failed to dock)
	Soyuz T-11	1984	7.9 d	Malyshev, Strekalov, Sharma
	Soyuz T-12	1984	11.8 d	Dzhanibekov, Savitskaya, Volk
	Soyuz T-14	1985	8.9 d	Grechko
Mir	Soyuz TM-3	1987	8.0 d	Viktorenko, Faris
	Soyuz TM-4	1987	8.0 d	Levchenko
	Soyuz TM-5	1988	9.8 d	Alexandrov[b], Solovyov, Savinykh
			9.0 d	Lyakhov
	Soyuz TM-6	1988	9.0 d	Mohmand
	Soyuz TM-7	1988	25.0 d	Chretien

[a] Rounded to the nearest tenth of an hour or day.
[b] Bulgarian cosmonaut.

TABLE 3
U.S. Manned Orbital Flights Lasting 28 Days or Longer[1,2]

Project	Mission	Launch year	Flight duration	Crew
Skylab	2	1973	28.0 d	Conrad, Kerwin, Weitz
	3	1973	59.5 d	Bean, Garriott, Lousma
	4	1973	84.1 d	Carr, E. Gibson, Pogue

TABLE 4
Soviet Manned Orbital Flights Lasting More Than 28 Days[1,2]

Project	Mission	Launch year	Flight duration	Crew
Salyut 4	Soyuz 17	1975	29.6 d	Gubarev, Grechko
	Soyuz 18	1975	63.0 d	Klimuk, Sevastyanov
Salyut 5	Soyuz 21	1976	49.3 d	Volynov, Zholobov
Salyut 6	Soyuz 26	1977	96.4 d	Romanenko, Grechko (1st prime crew)
	Soyuz 29	1978	139.6 d	Kovalyonok, Ivanchenkov (2nd prime crew)
	Soyuz 32	1979	175.0 d	Lyakhov, Ryumin (3rd prime crew)
	Soyuz 35	1980	184.8 d	Popov, Ryumin (4th prime crew)
	Soyuz T-4	1981	74.8 d	Kovalyonok, Savinykh (5th prime crew)
Salyut 7	Soyuz T-5	1982	211.3 d	Berezovoi, Lebedev (1st prime crew)
	Soyuz T-9	1983	149.4 d	Lyakhov, Aleksandrov (2nd prime crew)
	Soyuz T-10	1984	237.0 d	Kizim, Solovyov, Atkov (3rd prime crew)
	Soyuz T-13	1985	112.1 d	Dzhanibekov
			168.2 d	Savinykh (4th prime crew)
	Soyuz T-14	1985	64.9 d	Vasyutin, Volkov (5th prime crew)
Mir	Soyuz T-15	1986	125.0 d	Kizim, Solovyov (1st prime crew)
	Soyuz TM-2	1987	326.5 d	Romanenko
			174.7 d	Laveykin
	Soyuz TM-3	1987	160.3 d	A. Alexandrov
	Soyuz TM-4	1987	366.0 d	Titov, Maranov
	Soyuz TM-6	1988	239.0 d	Polyakhov
	Soyuz TM-7	1988	121.0 d	Volkov, Krikalyov

Soviet flights of fewer than 28 d are listed in Table 2. Soviet short-duration flights began with the six-flight Vostok program (1961-1963), followed by the two-flight Voskhod program (1964-1965).[1,2] The subsequent Soyuz program, initiated in 1967, provided a multipurpose spacecraft for transporting crews to and from orbiting space stations.[1] Soyuz vehicles have also been used for 12 independent manned missions. All missions to the Salyut 1 and 3 space stations lasted fewer than 28 d. Short missions to later space stations were generally to transport visiting crews.

U.S. experience with long-duration space flight is limited to the three Skylab missions of 1973-1974 (Table 3). In contrast, the Soviets have made many long-duration sojourns in space (Table 4).

III. PHYSIOLOGICAL RESPONSES TO SPACE FLIGHT AND SIMULATED WEIGHTLESSNESS

Altered function of the cardiovascular system with space flight probably has a pervasive influence on all of the body's systems. However, for this presentation, which has been limited to information directly related to the effects of weightlessness on tolerance to orthostatic stress, the discussion of physiological responses is based on a sequence of events hypothesized to occur upon exposure to weightlessness: fluid shifts; cardiovascular responses; fluid balance, electrolyte, and hormonal responses; and baroreceptor reflex function. A final section summarizes the evidence of impaired responses to orthostatic stress after exposure to actual and simulated weightlessness. Within each section, space-flight data precede data from simulated weightlessness. Space-flight data are presented historically, because each flight program built upon the knowledge and experience gained from earlier programs.

Data from simulation studies are separated according to studies that lasted 24 or fewer hours and those of longer duration, because during the first 24 h the cephalad fluid shift characteristic of weightlessness is nearly completed. Furthermore, such brief exposures have proven attractive to investigators who can measure a greater number of physiological variables concurrently.

Short-duration simulation studies have relied primarily on head-down tilt and water immersion. The variables that have been measured in these short-duration studies are listed in Table 5, and the physiological responses measured are summarized in Table 6.

Horizontal bedrest and bedrest at a head-down tilt angle of 4 to 6° are the most frequently used long-duration simulations of space flight. A few investigators, mostly Soviet, have immersed subjects in water for many days, a protocol that requires wrapping the subjects in a waterproof material to keep them dry. Table 7 lists the studies of long-duration simulated weightlessness that were reviewed and the variables measured in them. The physiological responses to these long-duration simulations of weightlessness are summarized in Table 8.

Crewmembers of both U.S. and Soviet missions have been between the ages of 28 and 59 years. Because the great majority have been men, it is not yet possible to delineate sex differences in adjustments of orthostatic function with space flight. We found no analyses of the correlation between age and the effects of space flight on orthostatic function. Subjects in both short- and long-duration simulation studies were usually men between 20 and 30 years of age.

Because space flight has involved a relatively small number of subjects, both statistically significant findings ($p < 0.05$) and suggested trends are presented. From the simulation studies, differences cited were statistically significant unless otherwise noted.

A. FLUID SHIFTS

The first response to the loss of the head-to-foot hydrostatic pressure gradient is physical: body fluids redistribute from the legs and lower trunk to the thorax and head until they reach a steady-state distribution, probably determined by a balance between vascular and tissue pressures. Manifestations of this cephalad fluid shift are among the earliest physiological effects of weightlessness. A variety of techniques have been used to document and quantify the fluid shift.

1. Space Flight
a. Signs and Symptoms
Symptoms and physical signs indicating a cephalad fluid shift in U.S. and Soviet crewmem-

TABLE 5
Summary of Studies Reviewed:
Short-Duration Simulations of Weightlessness (24 hours or less)

Control posture	Tilt angle (deg.)	Regional BV	Central BV	Heart size	Pulmonary BV	Thoracic fluid (Z_0)	CVP	Pulmonary art. P	Leg V	Cerebral BF (vel.)	SV/SI	CO/CI	HR/ECG	ABP	Urine	Total body water	Ref.
*	−30 / −40	X			X					X							31
Horizontal and Head-Down-Tilt Bedrest																	
Supine	−5						X		X		X	X	X	X	X		143
Supine	−5			X			X				X	X	X	X	X		36
Stand	−6												X	X	X		100
Supine	−5						X		X		X	X	X	X	X		32
Supine	−5								X				X	X	X		35
Sit	0			X	X												38
Supine	−6							X			X	X	X	X			68
Supine	−6							X			X	X	X	X			37
HDT	−6											X	X				132
Supine	−5				X		X		X		X	X	X	X	X		34
Supine	−6					X			X		X		X	X			33
HDT	−6					X			X				X	X			131
Supine	−5												X	X	X		67
Water Immersion																	
Sit															X		107
Sit			X												X		40
Sit															X		110
Semi-recline													X		X		105
*													X	X	X		103
Supine														X	X		106

TABLE 5 (continued)
Summary of Studies Reviewed:
Short-Duration Simulations of Weightlessness (24 hours or less)

Control posture	Tilt angle	Extracellular V	Interstitial F	Total BV	Plasma V	PRA	Aldosterone	ADH/AVP	ANF	SNS/Catecholamines	Baroreflex	LBNP	Head-up tilt	Stand test	Centrifuge	Exercise	Diurnal	Sex differences	Ref.
Horizontal and Head-Down-Tilt Bedrest																			
*	−30, −40																		31
Supine	−5			X		X						X				X			143
Supine	−5			X	X	X	X	X				X				X			36
Stand	−6					X	X	X											100
Supine	−5											X				X			32
Supine	−5	X	X																35
Sit	0																		38
Supine	−6									X									68
Supine	−6									X									37
HDT	−6											X				X			132
Supine	−5			X	X	X	X	X				X							34
Supine	−6											X							33
HDT	−6											X							131
Supine	−5				X	X	X												67
Water Immersion																			
Sit		X		X		X	X	X		X									107
Sit							X												40
Sit										X									110
Semi-recline		X	X	X	X	X		X					X						105
*															X				103
Supine			X		X														106

							Ref.
Stand						X	109
Sit	X		X		X		102
Sit	X	X					43
Sit	X	X					101
Stand							41
Stand		X					42
Sit		X	X		X		104

Bedrest and Water Immersion

					Ref.
5° HDT	X	X	X		69
*	X	X		X	133

Note: * = unspecified. For other reviews of responses to simulated weightlessness, see References 153—155. Abbreviations used in the table are as follows: ABP = arterial blood pressure; ADH = antidiuretic hormone; ANF = atrial natriuretic factor; AVP = arginine vasopressin; B = blood; CI = cardiac index; CO = cardiac output; CVP = central venous pressure; DBP = diastolic blood pressure; F = flow; HR = heart rate; LBNP = lower body negative pressure; MAP = mean arterial pressure; P = pressure; PRA = plasma renin activity; SBP = systolic blood pressure; SI = stroke index; SNS = sympathetic nervous system; SV = stroke volume; V = volume; Vel = velocity.

TABLE 6
Summary of Responses to Short-Duration Simulations of Weightlessness

Variable	Duration								
	0-15 min	30 min	30-60 min	90-120 min	3-4 h	6 h	8 h	16-20 h	24 h
Central BV	⇑40,101,41								
Right atrial V	⇑41								
Pulmonary BV	⇑31,34,38		⇓38	⇑36,⇓38					
Thoracic fluid (Z_0)	⇓33		⇓33	⇓33		-36			
Central venous P	⇑32,43,42	⇑36,32,34	⇓36,34	-34,⇑101	⇑32	-36			⇑34
Pulmonary artery P or RAP	⇑37			⇑34,46		⇑46			-34,⇑37
Leg V	⇓32,35,34,33	⇓35,33	⇓35,33	⇓35			⇓32		⇓34
Forearm B F	⇑31	⇑36,-37	⇑33,⇑43	⇑74	-32	⇑75		⇑32	⇑36,34,-37
Cerebral B F	⇑32,33	⇑36,33,-37	⇑43,104	-36	-32,⇑104				⇑36,-37
S V	⇑32,69			-36		-36			
CO				-36,⇑104,69		-36			
HR-rest	-32,⇓33,101,41	-37,⇓36,34,33	-33,⇓43	36,-35,⇓34,105,101	⇓43	-36,34	-35		-37,⇓36,34
SBP		-33	⇓67	⇓35	-35	-67	-35		
DBP		-33	⇓67	⇓35	-35	-67	-35		
MAP	-32	-36,33	-104,⇑33	-104	-104				-34
Urine F			⇑40,102,104	⇑104		⇑36,32,35,34	⇑35,40,105		
Urine Na+			⇑104	⇑40		⇑32,67	⇑40		
Total body water							⇓105		
Extracellular V							⇓35		
Interstitial fluid P			⇓106				⇓105		
Total B V						⇓34	⇓105		⇓36,32,34
Plasma V			⇓43	-104,43	-104,⇓43	-67,⇓32			⇓67
Hematocrit			-104,⇓43	⇓104	⇓104	-67			⇑67
PRA		⇓107	-67,⇓104	⇓104	⇓104				-34,⇑67
Aldosterone		⇑107	-67,⇓107,104	⇓104	⇓107			-32	-34,⇑67
ADH			⇓104,43	⇓104,43	⇓107,43				-34
ANF			⇑102		-107,110				
SNS									
Baroreflex function									
Orthostatic tolerance						⇓103	⇓103	⇓32	⇓34,103
G tolerance						-103	-103		-103
Max $\dot{V}O_2$								⇓32	⇓34

Note: Numbers in chart are reference numbers. ⇓ = decreased; ⇑ = increased; - = no change. Abbreviations are the same as in Table 5.

Table 7
Summary of Studies Reviewed:
Long-Duration Simulations of Weightlessness

Control posture	Tilt angle (deg.)	Duration (days)	Regional BV	Central BV	Heart size	Pulmonary BV	Thoracic fluid (Z_0)	CVP	Pulmonary art. P	Leg V	Cerebral BF (vel.)	SV/SI	CO/CI	HR/ECG	ABP	Urine	Ref.
Horizontal Bedrest																	
Supine	0	12		X													118
AMB	0	14–										X	X	X	X	X	73
AMB	0	21															81
AMB	0	15												X	X	X	134
	0	49												X	X		111
AMB	0	14		X								X	X	X			137
AMB	0	14												X	X		114
AMB	0	7												X			117
Supine	0	14													X		113
*	0	120												X	X	X	77
*	0	28												X			144
Sit/sup	0	20			X	X							X	X			70
AMB	0	15												X			120
AMB	0	28												X			82
AMB	0	48												X			112
Supine	0	14															78
AMB	0	14										X	X	X	X	X	145
AMB	0	14												X			146
	0	17															140
Head-Down Bedrest																	
Supine	–6	4												X		X	44
*	–4.5	120					X					X	X	X			76
*	–6	7												X			147
Supine	–4	30															138
AMB	–6	7												X	X		123
AMB	–6	30					X				X						148
Stand	–6	7											X			X	100
HDT	–6	2										X	X	X	X		45

(See Table 5 for abbreviations.)

Table 7 (continued)
Summary of Studies Reviewed:
Long-Duration Simulations of Weightlessness

Control posture	Tilt angle (deg.)	Duration (days)	Regional BV	Central BV	Heart size	Pulmonary BV	Thoracic fluid (Z_0)	CVP	Pulmonary art. P	Leg V	Cerebral BF (vel.)	SV/SI	CO/CI	HR/ECG	ABP	Urine	Ref.
AMB	−6	10												X			84
Supine	−6	30															79
*	−4	7									X						75
*	−4	182	X								X						74
AMB	−4	49												X	X		83
*	−15	7						X				X	X	X			80
*	−15	7						X	X			X	X	X	X		46
*	−4	30							X			X	X	X	X		71
*	−6	7								X				X			72
AMB	−6	7															136
AMB																	149

Horizontal and Head-Down Bedrest

Control posture	Tilt angle (deg.)	Duration (days)	Regional BV	Central BV	Heart size	Pulmonary BV	Thoracic fluid (Z_0)	CVP	Pulmonary art. P	Leg V	Cerebral BF (vel.)	SV/SI	CO/CI	HR/ECG	ABP	Urine	Ref.
AMB	−6,0	7												X			150
Supine	0,−4	5	X														139
Supine	−8,−12	0–7	X														151
AMB	−6,0	7			X							X	X	X	X		152
	−6,0	14											X				140

Water Immersion

Control posture	Tilt angle (deg.)	Duration (days)	Regional BV	Central BV	Heart size	Pulmonary BV	Thoracic fluid (Z_0)	CVP	Pulmonary art. P	Leg V	Cerebral BF (vel.)	SV/SI	CO/CI	HR/ECG	ABP	Urine	Ref.
*		7						X				X	X	X			121
*		7							X			X	X	X	X		47
*		7											X			X	85
*		28										X	X	X	X		142
*														X	X		141

(See Table 5 for abbreviations.)

Table 7 (continued)
Summary of Studies Reviewed:
Long-Duration Simulations of Weightlessness

Control posture	Tilt angle	Duration	Total body water	Extracellular V	Interstitial F	Total BV	Plasma V	PRA	Aldosterone	ADH/AVP	ANF	SNS/Catecholamines	Baroreflex	LBNP	Head-up tilt	Stand test	Centrifuge	Exercise	Diurnal	Sex differences	Ref.
Horizontal Bedrest																					
Supine	0	12					X								X						118
AMB	0	14					X	X					X									73
AMB	0	21				X	X											X			81
AMB	0	15				X	X								X			X			134
AMB	0	49	X				X														111
AMB	0	14				X	X								X		X	X			137
AMB	0	14					X							X				X			114
AMB	0	14					X							X							117
Supine	0	7														X					113
*	0	120															X				77
*	0	28		X																	144
Sit/sup	0	20	X				X										X	X			70
AMB	0	15		X		X	X							X							120
AMB	0	28					X							X	X	X		X			82
AMB	0	48					X								X						112
Supine	0	14					X											X			78
AMB	0	14	X	X			X			X				X	X		X	X			145
AMB	0	14					X	X		X				X	X		X	X	X	X	146
AMB	0	17																			X	140
Head-Down Bedrest																					
Supine	−6	4						X													44
*	−4.5	120						X	X	X											76
*	−6	7																			147
Supine	−4	30						X	X	X				X		X		X			138
AMB	−6	7				X	X	X						X		X		X			123
AMB	−6	30					X						X								148
Stand	−6	7										X									100
HDT	−6	2																	X		45
AMB	−6	10												X							84

Table 7 (continued)
Summary of Studies Reviewed:
Long-Duration Simulations of Weightlessness

Control posture	Tilt angle	Duration	Total body water	Extracellular V	Interstitial F	Total BV	Plasma V	PRA	Aldosterone	ADH/AVP	ANF	SNS/Catecholamines	Baroreflex	LBNP	Head-up tilt	Stand test	Centrifuge	Exercise	Diurnal	Sex differences	Ref.
Horizontal and Head-Down Bedrest																					
Supine	−6	30				×	×								×			×			79
*	−4	7																			75
*	−4	182																×			74
AMB	−4	49												×	×						83
*	−4	7						×	×					×				×			80
*	−15	7						×	×					×							46
*	−15	7												×							71
*	−4	30												×				×			72
AMB	−6	7																			136
AMB	−6	7																			149
AMB	−6,0	7												×				×			150
Supine	0,−4	5													×			×			139
Supine	−8,−12	0−7																			151
AMB	−6,0													×							152
	−6,0	14																		×	140
Water Immersion																					
*		7																			121
*		7					×	×												47
*		7															×				85
*																	×	×			142
*		28																×			141

Note: * = not specified. See Table 5 for abbreviations.

TABLE 8
Summary of Responses to Long-Duration Simulations of Weightlessness

Variable	\multicolumn{10}{c}{Duration (days)}									
	2	3-4	7	10-12	14-15	20-21	28-30	90	120	182
Central B V	⇓45					-73				
Pulmonary B V	⇓46					⇑70				
Thoracic fluid (Z_0)				⇓118						
Central venous P	⇓46		⇓46							
Pulmonary artery P	⇓46		⇓46							
Cerebral B F (vel.)	⇑74,45	-75								
LVEDV	⇓47		⇓47							
SV	⇓71,-45		-47			⇓70 -73	⇓72			
CO	⇓71,-45		-47			⇓70,-73				
HR — rest	⇓44,-45	⇓44,-148		⇑84,-148	⇑81,78,146	⇑70,-73	⇑72,-148			
HR — max					⇑146				-76	
SBP	-45				⇑146					
DBP	⇓45				⇑146,-73					
MAP		-148		-148	⇑114	-70	-148			
Urine F		⇓44								
Urine Na^+		⇓44	⇑85							
Urine K^+			⇓85							
Total body water			-47							
Total B V					⇓114	⇓70				
Plasma V		⇓148	⇓113	⇓118,148	⇓81,111,78,146, 73,113,-114	⇓70,73	⇓148			
PRA		⇑44	⇑85							
Aldosterone		⇑44	⇑85		⇑146	-73				
ADH		-44		-148		-73				
Norepinephrine		-148		⇓148	⇑146		-148			
Baroreflex function							⇓148			
Orthostatic tolerance					⇓114,120,146,113	⇓73	⇓82,113			
G Tolerance					⇓111,120,146,144,137			⇓138		⇓83
Max VO_2			-123,149		⇓81,78,70,146			⇓83		⇓83

Note: Numbers in chart are reference numbers. ⇓ = decreased; ⇑ = increased; - = no change from control. Abbreviations are the same as in Table 5.

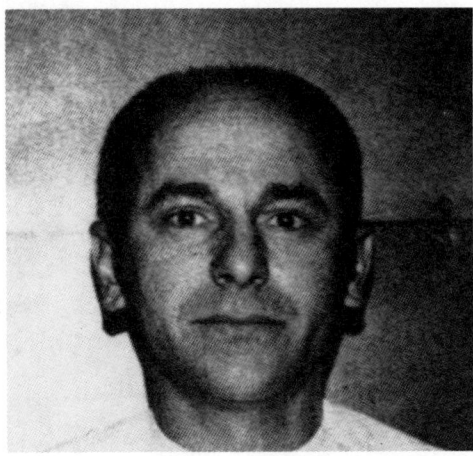

FIGURE 1.   Skylab astronaut on Earth (right) and during space flight. Note especially that during space flight his face has a puffy appearance, his eyes do not appear as deep set, and the creases around his mouth are gone. (NASA photograph from *Biomedical Results from Skylab.*)

bers have included fullness in the head (similar to the feeling of hanging upside down on Earth), nasal stuffiness, sinus congestion, rounding and redness of the face (Figure 1), puffiness of the eye sockets, redness of the eyes, engorgement of the veins of the head and neck, and a decrease in leg girth.[3-13] Unpublished reports by NSTS crewmembers suggest considerable variability in the type and intensity of their symptoms.

These signs and symptoms have characteristically appeared soon after orbital flight has been attained. In the early hours of orbital flight, head fullness and nasal stuffiness have been the most frequently reported sensations, and crewmembers have concurrently observed the development of facial roundness, head and neck vein engorgement, and decreased leg volume in each other.[3-7,14,15]

It was not until Skylab, however, that the duration of these signs and symptoms was appreciated. Photographs taken near the end of the 28-d and 59-d flights showed persistent facial puffiness.[8] Astronauts on the 84-d flight reported that puffiness of the eye sockets, facial rounding and redness, nasal and sinus congestion, and distention of neck and forehead veins lessened but never completely disappeared.[15] The eyes, however, gradually cleared. Only one of the three men on the crew felt as "normal" during the last 2 weeks of the mission as he had on the ground before flight. Signs and symptoms of the fluid shift improved for $\frac{1}{2}$ to 2 h after in-flight bicycle exercise, and one crewmember reported less head fullness after meals. Head fullness symptoms consistently felt worse at the end of the day. Calf circumference was restored by about $\frac{1}{2}$ in. during treadmill exercise but decreased again within 15 to 30 min after the exercise stopped.

Soviet cosmonauts have reported sensations of blood flowing or rushing into the head. Cosmonauts of early Soyuz flights noted that these sensations decreased during spacecraft rotation when they were oriented along the centripetal force vector with their heads toward the center of rotation.[9,10] Soviet reports about persistence of this symptom conflict.[10,11,16,17] For example, one report indicated that the blood-rushing sensation first appeared with the onset of weightlessness and disappeared after two or three orbital revolutions, while another indicated mild persistence of the symptom throughout a 17-d Soyuz mission.[10] Similar discrepancies exist in reports of later flights.[11,17] Confusion between the temporary sensation of fluid movement and the persistent feeling of head fullness may explain this conflict. After landing from a 63-d Salyut mission, one cosmonaut reported sensing fluid moving from the head back toward the feet.[11]

b. Limb Circumference and Volume Measurements

Measurements of calf circumference were made before and after most Apollo flights, and leg volume was estimated after two flights.[18,19,156] Two to eight hours after splashdown, after crewmembers (n = 24) had been standing and walking, the average calf circumference was 1.04 cm (2.8%) less than the last preflight value, a statistically significant decrease. For six Apollo crewmembers, the average volume of both legs after flight (estimated from multiple circumferences) was 1005 ml less than the last preflight volume, a highly significant difference.[18] By 48 h after splashdown, both calf circumference and total leg volume were still significantly less than the last preflight values. Moore and Thornton[20,21] speculated that the Apollo postflight measurements were made after most of the fluid had already returned from the upper body to the lower extremities. If so, the persistent reduction of calf circumference and leg volume after flight suggests several possibilities: (1) that more than a liter of fluid is shifted to the upper body in flight; (2) that because of the reduced body fluid volume, fluid is not available to shift back to the legs; or (3) that substantial leg tissue is lost.

Limb volumes were estimated before and after the first two Skylab missions, and in-flight measurements were added on the final Skylab mission.[8] At the first measurement, on the third day of flight, leg volume was reduced by about 1 liters per leg in the two crewmembers studied. Measurements on the eighth day showed a total loss from both legs of about 2 liters (about 13% of total leg volume) in all three crewmembers. Leg volume was restored nearly to preflight levels within several hours after landing, perhaps in part the result of a heavy in-flight exercise regimen. The crewmembers' arms showed no evidence of fluid shift.

During ASTP, the first in-flight leg volume measurement at 6 h was made on only one crewmember.[22] His left leg volume (only leg measured) decreased by 260 ml from the last preflight measurement. At 32 h, when all three U.S. crewmembers participated, left leg volume had decreased an average of 523 ml (range 350-860 ml). Left leg volume of two of the three crewmembers continued to decline throughout the flight. The lowest in-flight volumes, recorded after 7 d, averaged 817 ml less than the last preflight measurement. The first postflight measurements, 1.5 to 2.0 h after splashdown, indicated that an average 417 ml of left leg volume had been restored (from the last in-flight measurement). These data supported the speculation that the majority of the in-flight fluid shift occurred within the first day of flight.

For NSTS missions, Moore and Thornton[21] developed a stocking plethysmograph, which simplified the leg volume measurement procedure and improved the reproducibility of measurements (Figure 2). Preflight and postflight leg volume measurements with the stocking were made while crewmembers were standing. (The investigators noted that changing from the standing to the supine position in 1 G shifts about 300 ml of blood out of the legs.) During two NSTS missions in 1985, three crewmembers began in-flight leg volume measurements at 1 h, with four additional measurements during the first 10 h of flight and less frequent measurements during the remainder of the flights.[21] Relative to measurements made 4 h before launch, two crewmembers experienced their most rapid leg volume loss during the first hour of flight. In this hour, one crewmember lost about 500 ml (45%) of the total 1100 ml he lost from the measured leg during the mission. The other crewmember lost about 700 ml (78%) of his total 900-ml loss. Leg volume of the third crewmember was reduced at a constant rate for the first 2 h of flight; during this time, he lost about 850 ml (81%) of his 1050-ml total loss from the measured leg. The average loss of about 1 l per leg represented 12.6% of the leg volume measured 4 h before launch. At 1.5 h after touchdown, the three crewmembers had recovered 29, 75, and 82%, respectively, of their preflight (launch minus 4 h) volumes. Moore and Thornton noted that interpretation of data collected early in flight is complicated by the fact that crewmembers lie on their backs in the spacecraft with their legs elevated for a variable period of time prior to liftoff (Figure 3). Thus, fluid shifting may begin before launch, as suggested by astronaut complaints of bladder distention while still on the launch pad.

The second long-term crew of the Mir station experienced a 5 to 18% decrease of "shin"

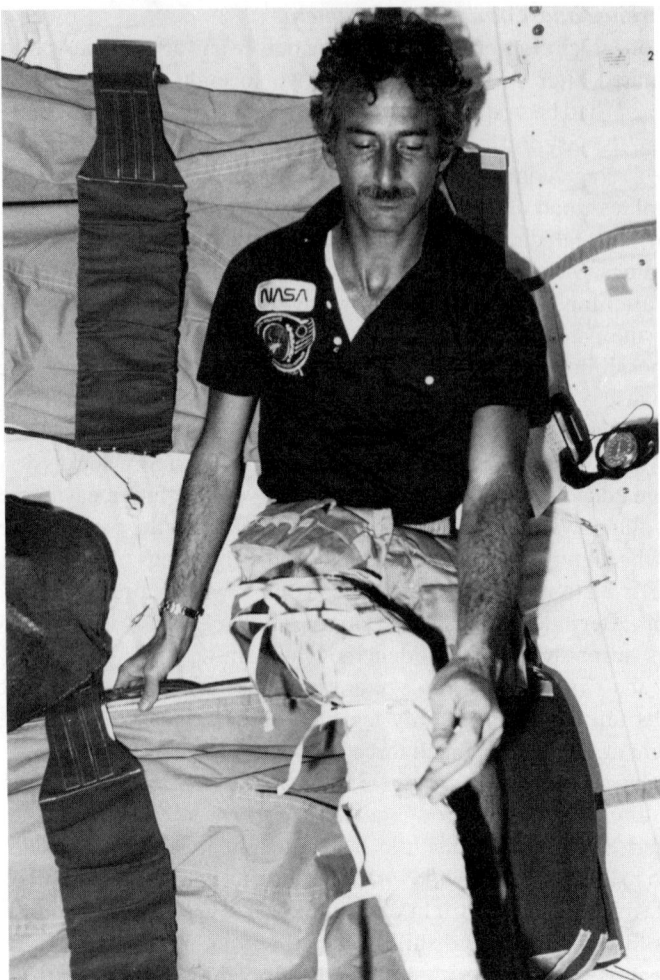

FIGURE 2. Stocking plethysmograph that has been used before, during, and
after flight on Space Shuttle to measure leg circumferences for the estimation
of changes in leg volume. (NASA photograph.)

volume during the mission and a 12% decrease after flight.[23] The timing of volume changes was
not reported.

c. Superficial Venous Blood Distribution

A series of preflight, in-flight, and postflight infrared photographs — front, side, and back
— made on crewmembers of the last Skylab mission showed qualitatively that leg veins were
not completely empty and that the jugular and head veins were always completely full and
distended in flight. After flight, reversion to the preflight pattern was reported to be prompt.[8]

d. Central Venous Pressure

Kirsch et al.[24,25] tested a premise of the hypothesis of Gauer and Henry[26] that the cephalad fluid
shift would increase central venous pressure during flight and initiate a negative water balance.
Eight crewmembers were tested during two NSTS Spacelab flights in 1983 and 1985. On the
ground, central venous pressure was determined by the method of Gauer and Sieker,[27] in which
subjects lie on their right sides with the right arm hanging vertically through a hole in the table.

FIGURE 3.   NSTS astronaut in prelaunch posture. Not shown are his legs, which are elevated with hips and knees bent at approximately 90 degrees. (NASA photograph.)

In flight, the antecubital vein pressure was assumed to be close to central venous pressure. Surprisingly, all in-flight central venous pressures, even those recorded at 20 to 40 min after launch in three subjects, were below preflight values, and in-flight hematocrits were elevated above preflight levels. Measurements from the 1985 flight showed that venous pressure continued to decrease over the next 3 d before stabilizing. At 1 to 4 h after flight, the eight astronauts had lost 4 to 9% of their body weight, which is consistent with either a negative water balance or a negative caloric balance. However, "central" venous pressures were higher and hematocrits lower than in-flight values, suggesting hemodilution. Twelve hours later, venous pressures were the lowest recorded during the study. The investigators concluded that a relocation of fluids stored in the extravascular compartment of the upper body in flight was already well underway in the first hour after landing, diluting   the blood and keeping venous pressure unexpectedly high. Based on the postflight data, the investigators proposed that cephalad fluid shifting in weightlessness had started quickly and lasted only several hours; any transient rise in venous pressure before this fluid moved to the upper-body extravascular space had apparently been missed by the earliest in-flight measurement. The supine position of the astronauts with their legs elevated for at least 2 h prior to liftoff could not be controlled. Furthermore, the comparability of the different techniques used in the gravity and weightless conditions is uncertain.

In 1986, one NSTS crewmember performed noninvasive central venous pressure estimations based on the method of Durr et al.[28] Jugular venous blood flow was monitored with a unidirectional Doppler flow detector while end-expiratory intrathoracic pressure was increased by partially occluded expiration. The intrathoracic (mouth) pressure that transiently interrupted jugular venous blood flow was the estimate of central venous pressure. In-flight values were consistently lower than the preflight average, decreased to their minimum value after 3 d, and then stabilized.[29]

Soviet researchers inferred that central venous pressure was increased during space flight because more negative lower body negative pressure (LBNP) was required during than before the flight to cause a similar change in jugular pulsations.[13] Their conclusions conflict with the more direct measurements of Kirsch et al.,[24,25] perhaps because the Soviet technique measures venous filling rather than venous pressure.

e. Impedance Measurements

Baisch et al.[30] recorded segmental bioelectric impedance in four crewmembers on the U.S.-German Spacelab D-1 mission. They defined two body segments (shoulder to waist [Z-torso] and shoulder to knee [Z-body]) and evaluated fluid volume using a body shell model. Baseline data were collected on the crewmembers 6 or 7 d before flight. Impedance was measured continuously for 38 h beginning 2 h before launch in two crewmembers, followed by two additional in-flight measurements on one of them. The other two crewmembers made single in-flight measurements. During flight, most Z-torso and Z-body impedance measurements were elevated above baseline. Estimated body fluid losses from the Z-body measurements were 1.3 to 2.0 l. The investigators concluded that most of the cephalad fluid shift took place while the astronauts were supine on the launch pad and that only fluid loss — and not fluid redistribution — could be inferred from their in-flight data.

In summary, the above data suggest that very early in flight, or even before launch, $1^1/_2$ to 2 l of fluid migrate from the legs to the upper venous compartment and are rapidly redistributed out of the venous compartment. By some as yet unspecified mechanism(s), body fluid is then reduced during flight. Although much of the central fluid shift occurs within the first hours of space flight, the reversal of this effect after landing is incomplete, even after several days.

2. Simulated Weightlessness

Fluid redistribution during simulated weightlessness mimics that during actual flight and supports the belief that the fluid shift begins with the head-down prelaunch posture. The data from simulated weightlessness studies suggest that body fluids shift immediately after the head-down posture is assumed and that the resulting increases in thoracic fluid and central blood volume reverse and return to normal 1-G levels within a day or two.

a. Studies Less than 24 Hours

Data from short-duration simulated weightlessness studies assist in understanding responses during two phases of space flight. First (from head-down tilt experiments), they mimic the physiological responses of crewmembers when they are in the head-down prelaunch posture (Figure 3) for up to 5 h. Second, they may provide valuable information about responses during the first hours of orbital flight, when obtaining flight data is most difficult.

i. Horizontal Bedrest and Head-Down-Tilt Bedrest

The orientation of subjects in bedrest has varied from horizontal to 45° head down, with most recent experiments using 5 or 6° head-down tilt. The process of fluid shifting toward the head may last for only 15 to 30 min or for as long as 2 h, depending on individual differences, the angle of head-down tilt, posture during the control period, and experimental interventions such as occlusion cuffs on the legs.

These initial fluid shifts have been documented by several techniques. Alekseyev et al.[31] measured regional blood volumes using impedance rheography and reported that the blood volume of the subjects' heads increased in the first minute of all their 30 and 40° head-down experiments. Reduced blood volume in subjects' legs immediately after the initiation of head-down bedrest was measured by reductions in leg circumference[32-34] and by fluid pressures measured with wick catheters.[35] Meanwhile, central venous pressure increased by approximately 2.5 cm H_2O in the first half hour of 5° head-down tilt.[32,34,36] Lollgen et al.[37] reported that right atrial pressure, pulmonary artery pressure, and pulmonary vascular resistance all increased during the first minutes of 6° head-down tilt. Pulmonary blood volume (measured by rheography) after 15 min of 30° head-down tilt[31] and pulmonary capillary blood volume (measured by carbon monoxide [CO] diffusion) after 90 min at 5° head-down tilt[34] and after 90 min horizontal bedrest[38] were increased. These changes were accompanied by increased pulmonary venous tone and decreased arteriolar tone.[31] A decrease in bioelectric impedance of the thorax after 10 min at 6° head-down tilt was further evidence for the shift of fluids into the upper body.[33]

After the first hour of bedrest, many of these initial indications of central fluid shift diminished or reversed. Central venous pressure returned to control levels within 60 min.[34] Pulmonary blood volume (measured by CO diffusion) decreased by 18.3%/h after the initial increase cited above, and after 90 minutes it was less than that in seated subjects.[38] Leg volume continued to decrease, but more slowly than at first.[35]

In our laboratory, we measured central venous pressure continuously via an indwelling catheter in subjects as they rested supine, moved to a standing position, and then assumed a recumbent posture with their hips and knees flexed 90° (mimicking prelaunch posture in the spacecraft).[39] The initial supine central venous pressure of 7.5 mm Hg fell to 4.5 mm Hg while subjects stood and then returned to approximately 7.5 mm Hg in the prelaunch position, but did not exceed these values, despite the greater fluid shift because the legs were raised. Thereafter, central venous pressure decreased continuously for the next 2 h and approached its value when subjects were standing (less than 6.0 mm Hg). These data indicate that central venous pressure of subjects in the prelaunch posture is no greater than it is when subjects are supine. If subjects are supine for an equilibration period before being tilted head down, their central venous pressure will have peaked and started to decrease. Then, with the head-down tilt and the additional fluid shift accompanying it, another transient increase would occur in central venous pressure, but not to values greater than those when the subjects were supine. These data further support the hypothesis that most of the cephalad fluid shift occurs before any measurements are made in flight and probably even before launch.

ii. Water Immersion

In this simulation of weightlessness, subjects usually sit immersed to their necks for 1 to 8 h in a tank of water (34.5 to 35.0°C). For control measurements, subjects sit in air, usually in an empty tank. The subjects stand periodically to urinate.

Pressures on the body during water immersion force fluid into the thorax. In the studies we reviewed, central blood volume[40] and central venous pressure increased immediately.[41-43] In one study by Risch et al.,[41] subjects' central venous pressure increased from –2.7 to +15 mm Hg. Norsk et al.[43] compared responses of subjects during 4 h of immersion to three different levels: umbilicus, midchest, and neck. Each immersion was performed on a different day. Central venous pressure was increased only with immersion to the chest or neck. Unlike the transient nature of increases in central venous pressure during head-down tilt and supine bedrest, the elevations were sustained for the duration of immersion. In this regard, head-down tilt and horizontal bedrest may be superior to water immersion as simulations of space flight. Therefore, we are concentrating on bedrest and head-down tilt, with less emphasis on water-immersion experiments.

The reasons for this difference have not been clearly defined. We suggest they may be related to the continued pressure on the lower body and abdominal organs, which reduces their capacitance, and to the hydrostatic pressure gradient between the head and the central circulation. Greenleaf (personal communication) has suggested that the fluid shift following bedrest and weightlessness may come mainly from the vascular compartment, with some contribution from the intracellular fluid, whereas the fluid shifted during immersion comes mainly from the intracellular compartment.

b. Studies Longer than 24 Hours

The studies we reviewed that evaluated fluid, cardiovascular, and related physiological responses of human subjects to horizontal bedrest of longer than 24 h ranged in duration from 10 to 120 d. Studies using head-down tilt bedrest ranged from 2 to 182 d.

i. Horizontal and Head-Down-Tilt Bedrest

Both of these simulations induce fluid shifts similar to those during weightlessness, but some conflicts exist. Eight subjects of Hyatt and West each performed two 1-week periods of bedrest: one horizontal and one head down.[43a] Although the subjects experienced an initial sensation of fullness of the head during 5° head-down tilt, they reportedly lost the sensation quickly. On the other hand, head congestion and headache persisted after 4 d of 6° head-down tilt for eight subjects studied by Annat et al.[44]

Changes in other variables measured during bedrest also mimic the corresponding changes during space flight. In the study by Hyatt and West, leg volume of the subjects, which was estimated from three conical sections extending from the foot to the thigh, decreased during both the horizontal and the head-down bedrest periods. The reduction during head-down bedrest was not significantly ($p < 0.10$) greater than that during horizontal bedrest. In another study, the thoracic impedance of nine men decreased immediately after the subjects were tilted 10° head-down, but returned toward pretilt, seated values by 48 h.[45] These data indicated a loss of fluid from the thorax after 2 d of head-down tilt. Katkov et al.[46] found that the central venous pressure of subjects tilted 15° head-down was lower than that of supine subjects by the second or third day, and it remained lower for 7 d.

B. CARDIOVASCULAR RESPONSES

Although elevation of central venous pressure has not yet been demonstrated during space flight, the available in-flight bioelectric impedance and venous pressure data and regional blood distribution data from simulation studies suggest that fluid moves rapidly toward the head during the head-down prelaunch period. The major pathway of the shift is probably the venous system. Thus, we present the immediate and long-term responses of the cardiovascular system before considering reflexive neurological, fluid and electrolyte, and endocrine changes.

1. Space Flight

The complexity of the early piloted missions limited the in-flight study of cardiovascular variables to the telemetric monitoring of the electrocardiogram (ECG) and occasional noninvasive measurements of arterial blood pressure. During the second decade of space flight, more detailed cardiovascular studies of cardiac and peripheral vascular structure and function were conducted.

a. Heart Rate

Key periods of flight-related stress were identified early in the American and Soviet space programs by heart rate monitoring. With respect to the development of orthostatic dysfunction, questions of interest concern (1) the relationship between heart rate and flight duration and (2) whether heart rate differs during and after the flight from preflight values.

Heart rates measured in resting crewmembers have been consistently higher during and after flight than before flight. Maximal heart rates occurred at launch, with entry into orbit, with extravehicular activity, and on return to Earth.[3,9,48-50] The stress tachycardia of launch and orbital entry prevents any evaluation of crewmembers' immediate heart rate responses to the cephalad fluid shift. However, in-flight heart rates during two Mercury missions were 13 and 24% higher than preflight averages.[14,51] Furthermore, all three astronauts on the final Skylab mission had higher heart rates when they were resting during flight than they had before or after flight, with the on-orbit rates of each crewmember typically exceeding the upper 95% confidence limit of his average preflight rate. Soviet reports indicate that on short flights heart rates have generally stabilized near preflight levels.[9,16,49,50] Data from several Soviet long-duration missions, however, suggest a tendency for heart rate to increase with flight duration. The resting heart rates of six cosmonauts on Salyut 6 (96 to 175 d) were generally at or above their preflight levels,[52] and heart rates of seven cosmonauts on Salyut 7 (149 to 237 d) tended to increase moderately.[53] However, although the heart rates of both cosmonauts of a 30-d mission on Salyut 4 were higher than preflight levels, they did not increase with time,[17] and, during the following 63-d mission, one crewmember maintained heart rates above preflight levels, while the other maintained rates below preflight levels.[17]

Heart rates measured after landing have consistently been higher than those measured before U.S. space flights, in part as a compensatory mechanism to altered orthostatic function. However, emotional stress, dehydration, and the warm temperatures at U.S. landing sites have probably contributed to the higher rates. At rest after landing, the pilots of two Mercury flights had heart rates that were 22 and 15% higher than preflight values.[14,48] The average heart rates of all Gemini astronauts as they rested after splashdown were 18 to 62% greater than preflight rates,[4] and heart rates of 24 Apollo astronauts when they were supine were approximately 15% higher after flight than before.[19] After the first four NSTS flights, heart rates were 20% greater than before flights after 2-d missions, 47% greater after an 8-d mission, and 26% greater after a 7-d mission.[54] After NSTS flights STS-5 to -8, heart rates were 31% greater than they were before flight, despite ingestion of salt and water by the crewmembers before reentry.[55] These data suggest that the decrease in body fluids during space flight is not the only cause of postflight cardiovascular stress.

b. Blood Pressure

Arterial blood pressure during space flight has changed less — and less consistently — than heart rate has changed. Although systolic pressures of two Mercury pilots were elevated 13 and 6% above preflight values, their diastolic pressure increases were only 7 and 3%,[48] and in-flight blood pressures of crews of three Gemini missions were interpreted as normal and relatively stable.[3] Furthermore, although all three Skylab 4 crewmembers had elevated heart rates on orbit, no consistent difference between their preflight and on-orbit blood pressures was reported.[56] In-flight arterial blood pressures of one crewmember during the STS-8 mission remained within a much narrower range than they did during normal activities on Earth.[56a]

The Soviets have reported blood pressures measured during several flights. Preflight and in-flight measurements by a physician cosmonaut on his two colleagues during a 1-d mission showed an initial decrease in both systolic and diastolic pressures and then an increase; pulse pressure was decreased.[50] During an 18-d mission, measurements taken during rest were remarkably consistent, with only a 5-mmHg range in systolic values and a 5- to 10-mmHg range in diastolic values.[16] The relationship between preflight and in-flight blood pressure levels was not reported.

Blood pressures after early missions appeared to be low[51] or did not differ significantly from preflight measurements.[19] After NSTS flights, however, mean arterial blood pressures of crewmembers measured when they were supine after flight have been significantly higher (by 14 to 18%) than those measured before flight.[57] Most of these crewmembers consumed salt and water before reentry. The blood pressure response to standing has depended on whether the crewmembers ingested salt tablets and water prior to entry and landing. If so, the postflight response has been a slight increase (2 mmHg) in mean pressure; if not, it has been a slight decrease (7 mmHg).

c. Cardiac Function

Echocardiographic analysis of cardiac function began with the last Skylab flight. M-mode echocardiography was performed before and after flight on the three astronauts who were first supine and rolled slightly onto their left sides (left lateral decubitus position) and then to the supine position for LBNP tests.[58] The left lateral decubitus position has been standard for most of the echocardiographic examinations. Relative to preflight data when crewmembers were resting, left ventricular end-diastolic volume (LVEDV) and stroke volume were reduced immediately after flight. Reductions were greatest in the two crewmembers whose stroke volumes were above normal before flight and persisted for more than 10 d in these two crewmembers. However, plots of LVEDV vs. stroke volume during the LBNP tests before and after flight were described by virtually identical linear regression equations, suggesting no detectable alteration in the usual Frank-Starling relationship. The observed reduction in LVEDV was consistent with the reductions in red cell mass and plasma volume that all three crewmembers experienced during the flight.[56] Preflight and postflight determinations of left ventricular wall thickness for all three Skylab 4 crewmembers were within the normal range and did not differ after flight.[58]

The U.S. resumed echocardiographic measurements before and after NSTS flights. Bungo et al.[55] reported data from two-dimensional echocardiograms on 17 crewmembers of NSTS flights with mission durations of 5 to 8 d. Postflight data were obtained within several hours of landing (day L+0) for seven of the crewmembers. LVEDV and stroke volume indices (adjusted for body surface area) were decreased significantly from preflight values (by 23 and 29%, respectively); heart rate, mean arterial pressure, and calculated systemic vascular resistance were elevated significantly (by 31, 11, and 31%, respectively). The left ventricular end-systolic volume (LVESV) index (7% decrease), cardiac index (11% decrease), and ejection fraction (9% decrease) were not significantly changed, but the left ventricular stroke work (in g × m) decreased significantly (by 22%). (The percent changes we cite here were computed from data in the original report.) All astronauts ingested salt tablets with 1 liter of water in the hour before re-entry. In a subsequent analysis of the data from 19 NSTS crewmembers,[59] a comparison was made between the six with the smallest LVEDV indices preflight and the six with the largest. Consistent with Skylab data, only the latter subgroup showed significantly reduced LVEDV and stroke volume indices after flight, but heart rate and mean arterial pressure were elevated significantly in both subgroups after flight. Total peripheral resistance was greater in the subgroup with the smaller LVEDV and stroke volume indices, both before and after flight.

At'kov reported preflight and postflight echocardiographic data from Soviet cosmonauts who spent 75 to 237 d in orbit.[60,61] LVEDV was reduced by 17% and stroke volume by 12% after flight. Concurrently, LVESV was reduced by 23%, ejection fraction was increased by 7%, and the velocity of circumferential shortening was increased by 9% (our computation from the data reported). No statistical comparisons of preflight vs. postflight data were reported. At'kov concluded that although left ventricular volume and stroke volume were reduced, contractility was not changed. Echocardiographic parameters were reported to return toward preflight values by 1 week after flight, and all parameters were reported as "normalized" by day 25 after flight.

To date, in-flight echocardiographic data have been reported for seven crewmembers on short-duration flights. In 1982, the French crewmember Jean-Loup Chretien performed echocardiographic studies on himself during a 7-d mission aboard Salyut 7.[62] In 1985 echocardiography was performed on two additional crewmembers aboard a 7-d U.S. NSTS mission[63] and on four crewmembers on another 7-d NSTS flight.[64] Left ventricular volumes, stroke volumes, and cardiac output increased slightly on the first day in flight compared with preflight values. Left ventricular and stroke volumes then decreased to 15% less than preflight values for the remainder of the flights,[63,64] and cardiac output returned toward preflight values.

Stroke volumes were estimated by vibrocardiography before and after three Apollo flights, and the preflight-to-postflight reduction calculated by this method was consistent with echocardiographic data from Skylab 4.[19] Compared with the average of three preflight measurements, stroke volume for the nine Apollo crewmembers was reduced by 14% at the first postflight evaluation.

d. Heart Size and Structure

Heart size may be estimated from the cardiothoracic ratios computed from standard posterior-anterior chest X-rays. These were taken at each crewmember's last major preflight medical examination and at his first postflight examination.[156] The average decrease among Apollo crewmen was 5% and was highly significant.[18,19] The pooled data from Mercury, Gemini, and Apollo crewmen showed that the peak decrement in the cardiothoracic ratio occurred after flights between 100 and 200 h.[19] Furthermore, the cardiothoracic ratios of crewmembers who walked on the lunar surface were unchanged after flight, in contrast to the decrement that was observed in Apollo crewmembers who did not. A limitation of most of these data is that the ECG trigger for synchronizing X-ray exposures with the end of systole and diastole was not used until the final two Apollo missions.

No consistent findings with respect to the cardiothoracic ratio were apparent after the ASTP, but these evaluations may have been complicated by the postflight pulmonary involvement caused by toxic gas inhalation.[65]

Heart size of eight of the Skylab crewmen was determined before and after flight from X-ray exposures, synchronized with the end of systole and diastole, and by incorporating a geometrical measuring system to compensate for changes in the subjects' positions or inspiratory levels.[66] These data revealed no roentgenological abnormalities, but both the cardiothoracic area ratio and the cardiothoracic diameter ratio were decreased significantly after flight, compared with preflight data. Furthermore, the postflight decrease in heart size was significantly correlated (0.91) with the augmentation of the heart rate response to LBNP.

e. Venous Function

Immediately after flight, when the Mercury astronauts stood or were tilted upright, their leg veins were engorged and their feet and legs took on a dusky reddish color.[14,51] Gemini astronauts' calf circumferences increased 12 to 82% more during postflight tilt-table tests than they had during preflight tests.[4] Maximum calf circumferences of 24 Apollo crewmembers were measured while they were supine before LBNP, and the percentage increase in calf volume during LBNP was computed. In contrast to the Gemini tilt-table data, the percentage increase in calf volume during LBNP was less in postflight tests than it had been before flight. Hoffler et al.[18] suggested that the difference between Gemini and Apollo findings was related to the testing technique. Changes in leg circumference could be monitored more faithfully during LBNP than during head-up tilt, during which the legs may move. Furthermore, a double-strand Hg-in-silastic gauge was used during the Apollo tests, rather than the single-strand gauge used previously. During the flight of Skylab 4, increases in calf volume during LBNP were substantially greater than those measured before flight or after flight (Figure 4).[56]

2. Simulated Weightlessness
a. Studies Less than 24 Hours

In the first hour of horizontal or head-down bedrest, cardiovascular variables respond to the

FIGURE 4.   In-flight LBNP testing aboard the second Skylab mission. (NASA
photograph.)

initial cephalad fluid shift, for instance, heart rate decreased in several studies.[33,34,36] Tomaselli
et al.[33] reported that the mean arterial pressure was increased by 45 min of 6° head-down tilt, but
Volicer et al.[67] observed decreased systolic and diastolic pressures after 1 h of 5° head-down tilt.
Knitelius and Stegemann[68] calculated the heart volume of nine men from longitudinal,
transverse, and anterior-posterior dimensions on frontal and lateral X-rays. The subjects' total
heart volumes averaged 746 ± 104 ml when they were prone horizontal and increased to 785 ±
125 ml (5.2%, $p < 0.025$) when they were tilted 6° head down for 5 min.

As a result of the increased cardiac preload, increases in stroke volume and cardiac output
have been observed, both by impedance cardiography[33] and by the acetylene rebreathing
technique.[32] However, some other studies using the acetylene technique either failed to show
increases in stroke volume and cardiac output,[34] or the increase in stroke volume was offset by
a decrease in heart rate.[36]

After the first 30 to 60 min of bedrest, many of these initial cardiovascular changes either
diminished or reversed. Heart rate returned to pretilt levels by 6 h of 5° head-down tilt.[34,36] The
investigators who reported that arterial pressures decreased early in head-down tilt reported that
they had returned to control levels by 6 h.[67] By 1 h, stroke volume and cardiac output of subjects

tilted head down were decreased below values for supine subjects,[33,36] and LVEDV peaked at 90 min.[36]

i. Water Immersion

The fluid shifts induced by water immersion also elicited cardiovascular changes. Heart rate decreased from a control value of 82 bpm in subjects seated in air to 65 bpm during immersion.[41] Unlike the transient decrease that occurs with head-down tilt, the decrease in heart rate with immersion — like the increase in central venous pressure — is sustained. Risch et al.[41] measured heart size — especially the right atria — by roentgen cinematography and observed a 31% increase within six heart beats after the subjects were immersed in water. This increase is larger than that observed by Knitelius and Stegemann[68] because of differences in technique (Knitelius and Stegemann measured volume of the whole heart) as well as differences in the stresses. Risch et al. also observed an increase in central venous pressure immediately upon water immersion.[41] With immersion to different levels of the body, stroke volume, cardiac output, and systolic pressure increased proportionately to the level of immersion, while heart rate and peripheral resistance decreased.[43]

Bonde-Petersen et al.[69] subjected four men to both head-down tilt and water immersion. Heart rates were the same in both simulations of weightlessness, but blood pressures were lower during water immersion than during head-down tilt.

These cardiovascular data from the first moments of simulated weightlessness must be related to flight data with caution, because there are unique physical stresses, e.g., acceleration stress, as well as emotional stresses during the first hours of space flight.

b. Studies Longer than 24 Hours
i. Horizontal and Head-Down-Tilt Bedrest

As during space flight, cardiovascular function changes and exercise capacity is reduced during long-duration bedrest. Saltin et al.[70] placed five men in bed for 20 d. Two of the men were classified as highly physically fit and three as moderately fit. Subjects were allowed to walk to the bathroom and to sit when they ate or read. After 20 d, their basal heart rates had increased by approximately 9%, while their blood pressures and total peripheral resistance were not changed. Heart volume, plasma volume, and red cell mass were reduced by 11, 9, and 6%, respectively. The most striking effect of bedrest was the impairment of circulatory response to exercise, as evidenced by a 28% reduction in maximum oxygen uptake.

Annat et al.[44] reported that the average heart rate increased after 4 d of 6° head-down tilt, and stroke index and cardiac index were decreased by the second or third day of 15° head-down tilt in a study by Katkov et al.[71] Similarly, Maksimov et al.[72] found that stroke volume and cardiac output (measured by impedance cardiography) increased in their subjects during the first week of head-down tilt. Total peripheral resistance decreased. These changes reversed by 30 d.

On the other hand, Frey et al.[45] found that average heart rate, stroke volume, and cardiac output (measured by impedance cardiography) of nine subjects were not changed after 2 d at 10° head-down tilt; but their arterial diastolic pressure was 8% greater than that measured within the first 2 h after head-down tilt began. Chobanian et al.[73] monitored subjects when they were resting before and after 3 weeks of horizontal bedrest and found no changes in heart rate, cardiac index (measured by dye dilution), left ventricular ejection time index, mean arterial pressure, peripheral resistance, or central blood volume after bedrest.

Other cardiovascular variables that may influence orthostatic tolerance are cerebral blood flow and electrocardiographic changes. Guell et al.[74] calculated cerebral vascular resistance from Doppler flow velocities measured in the common and internal carotid arteries. They observed that cerebral vascular resistance of four men increased after they were tilted 4° head down. Flow velocities decreased after 2 h but then increased after 48 h. In another study, these investigators found that regional cerebral blood flow of six men was increased after 6 to 9 h of

4° head-down tilt but returned to control levels by 72 h.[75] This increase in blood flow was accompanied by fullness in the head, nasal congestion, and edema. Frey et al.[45] observed that blood velocity in the middle cerebral arteries of nine men was slower ($p < 0.06$) after 48 h of 10° head-down tilt than it had been immediately after head-down tilt was initiated. Subjects studied by Artamonova et al.[76] exhibited no changes in resting heart rate, conduction time, or amplitude of the P or QRS waves or of the ST segment of the ECG after 120 d of 4.5° head-down tilt. However, these subjects had some T-wave changes that were similar to those seen in ECGs of cosmonauts after flight: the waves were flatter and wider, and some had two peaks. Decreases in serum K^+ concentration and increases in Ca^{2+}, total calcium, and magnesium concentrations all correlated significantly with changes in the T waves of the ECGs.

Occasionally, especially in very long bedrest studies, cardiovascular variables oscillate for unknown reasons. In one such study, the caliber of the retinal arteries and veins (measured by photoprojection) and the intraocular pressure of 10 men increased initially and then oscillated throughout the 120-d period of bedrest.[77]

Frey et al.[45] found that the morning-to-evening variation in several cardiovascular variables was greater than the changes after 2 d of head-down tilt. Heart rate, stroke volume, cardiac output, and blood velocity in the middle cerebral artery were greater in the evening than in the morning; total peripheral resistance tended to be higher in the morning ($p < 0.07$); and systolic arterial pressure tended to be higher in the evening ($p < 0.07$). They concluded that the times of measurement must be carefully considered when evaluating the effects of real or simulated weightlessness.

Exercise and other countermeasures have been examined for their effectiveness in restoring aerobic capacity, which is usually decreased after periods of horizontal or head-down-tilt bedrest. If intensive dynamic ergometer exercise is performed daily during bedrest, the loss of aerobic capacity is reduced or prevented.[78-80] Performance of isometric or isokinetic exercise[78,79] or the use of a "reverse-gradient" garment[81] during the bedrest period has also reduced the loss of aerobic capacity, but neither occlusion cuffs around the thighs nor administration of 9-alphafluorohydrocortisone attenuated the postbedrest loss of aerobic capacity.[82] Kakurin et al.[83] reported that subjects who used a battery of countermeasures, including exercise, LBNP, electrical stimulation of muscles, and saline ingestion, did not experience a loss of aerobic capacity after 182 d of 4° head-down tilt; whereas aerobic capacity of their subjects who did not receive the countermeasures decreased by about 40%.

Some investigators have used mathematical analyses to extract more physiological information from their data. Goldberger et al.[84] performed spectral analysis of the heart rates of 10 men, 35 to 49 years of age, who were monitored before and after 10 d of 6° head-down bedrest, during graded LBNP to –100 mm Hg, and before and after administration of atropine. The subjects' heart rates at rest were higher after bedrest, but spectral analysis revealed no differences in variability. When atropine was administered to the subjects after bedrest, their heart rates increased further; spectral analysis indicated a decrease in variability, which the authors attributed to a reduction in heart-rate reserve. The duration of tolerance to graded LBNP, which was greater before the bedrest than after when the subjects received a placebo, did not differ between before and after bedrest when the subjects were given atropine. However, the variability in interbeat interval decreased after bedrest. These investigators concluded that bedrest induces an instability in the sympathetic nervous system control of the heart rate.

ii. Water Immersion

After 7 d of water immersion, subjects studied by Shulzhenko et al.[85] lost 1073 ml fluid. Stroke volume and cardiac output (estimated by rheography), which had increased transiently on the first day, were reduced to control levels, while total peripheral resistance was elevated.

While horizontal and head-down-tilt bedrest simulate the fluid shifts of space flight reasonably well, their relationship to in-flight cardiovascular changes is more difficult to evaluate. One complication is the lack of agreement among the data from simulation studies. In some, heart rate and arterial pressure increase and cardiac output decreases with bedrest; in others, no differences exist from prebedrest values. Many of these differences probably result from differences in experimental design. For example, investigators have used different tilt angles and different rules governing subject movement and head elevation; different measurement techniques (sometimes invasive, sometimes noninvasive), with some techniques more reliable than others; and different numbers of subjects, such that a response might not be significant in a study with too few subjects. A second problem is that much of the space-flight data have been obtained before and after flight only. After flight, variables such as temperature at the landing site and salt and water consumption almost certainly obscure flight effects. A third complication is the time of day when data are recorded. Most cardiovascular variables conform to diurnal variations. Thus, repeat measurements and measurements to be compared between flight and simulations should be measured at the same time of the astronaut's "day". Failing that, the time of day should be noted so data can be corrected and compared properly.

Despite these complications, however, horizontal bedrest and head-down-tilt bedrest seem to be reasonably good simulations of space flight for studying the cardiovascular changes and have great potential value. Those simulation studies that do show changes show them in the same direction as changes with space flight: for instance, heart rate increases and cardiac output decreases. These bedrest simulations allow us to test interventions such as exercise and LBNP that are potential in-flight countermeasures to deconditioning.

C. OTHER PHYSIOLOGICAL ADJUSTMENTS TO FLUID SHIFTS

Knowledge about fluid-shift responses on Earth suggests that the rapid relocation of fluid toward the head in weightlessness should stimulate body fluid-control mechanisms to respond to a perceived fluid overload. After that, long-term adjustments to the redistribution of body fluid are made. In this section, the evidence for immediate and long-term changes in fluid and electrolyte balance in actual and simulated weightlessness is summarized. Data concerning the endocrine mechanisms that affect these changes and may be reset with exposure to weightlessness are also summarized.

1. Space Flight
a. Body Mass

Leach[86] reported that net (preflight-to-postflight) weight losses of crewmembers of the Mercury, Gemini, and some Apollo missions (0.2 to 14 d) ranged from 0.8 to 9.1% of preflight weights. Using her data, we found no statistically significant linear correlation ($r = 0.29$) between flight duration and the percent loss of body weight. Soviet cosmonauts have also experienced net weight losses during short-duration missions. According to Leach's summary[86] of all the Vostok and Voskhod missions (0.1 to 5 d duration), these cosmonauts lost 0.2 to 4.0% of their body weight. As with the U.S. astronaut data, percent weight loss was not correlated with flight duration. Measuring body mass during flight for evaluating the time course of weight loss has been possible only during space-station missions. Aboard all three U.S. Skylab missions, crewmembers used a spring-mass oscillator for daily body mass measurements.[87] Mass was consistently lost during the first 5 to 10 d of flight, suggesting the loss was mainly fluid. Later during flight, additional mass was lost on Skylab 2, but mass was recovered by some crewmen on Skylab 3 and 4 as caloric intake was increased.

b. Fluid Balance

An important question regarding the mechanisms of space-flight orthostatic dysfunction is whether the early cephalad fluid shift initiates a diuresis in astronauts after attaining weightlessness. While this diuresis has been predicted on the basis of the Henry-Gauer reflex, it remains undocumented. Data from the Mercury MA-8 astronaut suggested both a diuresis that began on the launch pad and a net fluid loss during flight.[14] This astronaut voided three times into the urine collection device during the 155 min he sat in the spacecraft prior to launch, despite having drunk only 325 ml of fluid from the time he awoke until launch (5.5 h later). During flight, he drank 500 ml of fluid and voided three times. Total urine output was unknown, but the loss of 2.05 kg during the 9.2-h flight suggested a substantial negative fluid balance caused in part from sweating as a result of elevated ambient temperatures. During the first 24 h after flight, this astronaut drank 2580 ml of fluid but had a urinary output of only 775 ml. On the other hand, diuresis was not evident during the 34-h Mercury MA-9 mission, although the astronaut consumed approximately 1500 ml of water and lost almost 3.64 kg during the flight.[51] In the first 24 h after landing, his fluid intake was 3900 ml, while his urinary output was only 545 ml.

During the 14-d Gemini flight, urine volume was estimated from the flowmeter on the urine collection device, and urine samples were retained for measurement of total volume using a tritium-dilution technique.[3] Fluid intake was the lowest for both crewmembers during the first 24 h of flight (about 300 ml and 500 ml). Urinary output was less than 150 ml for one and 0 ml for the other astronaut during the second 24 h, but returned to a more normal level thereafter. Immediately after this 14-d flight, fluid was retained and weight was gained by the crewmen.[88] Apollo crewmembers lost an average of 3.51 kg (5% of preflight weight), approximately one third of which was regained in the first 24 h after landing.[89]

Observations from 20 cosmonauts before and after short (4 to 14 d) flights have been similar to those from U.S. astronauts, that is, positive fluid balance after landing, with reduced urinary output and increased water consumption.[90]

The Skylab program provided fluid balance data before, during, and after long-duration space flight. During the first 6 d of flight, the nine crewmen were in negative water balance. Available urinary output data did not reflect a diuresis; decreased intake and increased insensible water loss because of the low ambient humidity probably account for this negative water balance. The average daily urine output of the crew was 400 ml, which was less than their output before flight; and their fluid consumption was 700 ml less than their average consumption before flight.

Total body water of three male NSTS crewmembers was measured using an isotope-dilution technique using water with ^{18}O before flight; on days 2, 4, and 5 of the flight; and after the flight.[92] Total body water decreased by 3% from preflight to flight day 2 ($p < 0.10$); on days 4 and 5, it did not differ from day 2, and postflight values did not differ from those during the flight.

Thus, a negative fluid balance during space flight has been indicated. Its causes may include reduced fluid intake and increased excretion by nonurinary routes, resulting partly from the dry and, in early flights, hypobaric cabin environment — but probably not in-flight diuresis. Diuresis before launch, however, may affect fluid-balance loss measurements made early in flight.

c. Plasma and Blood Volume

Plasma volume was measured using the radioiodinated serum albumin dilution technique before and immediately after several Gemini flights, and blood volume was calculated from plasma volume and hematocrit.[3] Plasma volume was reduced by 13% after the 4-d flight, by 6% after the 8-d flight, and by 10% after the 14-d flight. Average blood volume was reduced by 10.5% after the 4-d flight and 13% after the 8-d flight, but was unchanged after the 14-d flight. Thus, red blood cell mass was unchanged after 4 d but was reduced after 8 and 14 d. Nonsignificant decreases in plasma volume, extracellular water, and total body water were measured in two of the ASTP crewmen after flight, but these data should be interpreted with caution because of possible complications resulting from exposure of the crew to hazardous fuels.[93] Johnson et al.[94] used radioisotope techniques to measure plasma volume, red blood cell mass, and red cell life span before and after Skylab flights. Compared with preflight levels, plasma volume was decreased by 8.4% immediately after the 28-d mission, by 13.1% after the 59-d mission, and by 15.9% after the 84-d mission. Red cell mass (determined with[51]Cr tagging) was reduced significantly — from 2075 ml before flight to 1843 ml after flight. (Red cell mass of a control group measured at the same times and places was unchanged after the flight.) Since life-span characteristics of the crewmen's red blood cells did not differ before and after flight, the loss of red blood cell mass was probably not the result of hemolysis. The investigators concluded the cause was reduced bone-marrow activity.

d. Electrolytes and Hormones
i. Urine

Urine was collected for analyses before, during, and after the 14-d Gemini flight. Compared with preflight values, urinary K^+ and Na^+ were reduced during and after flight, and aldosterone[4] and catecholamines were increased during flight.[86] Urine (and plasma) samples were collected before flight and on the landing day after the Apollo missions. Urine levels of aldosterone and antidiuretic hormone were elevated significantly (by 57 and 152%, respectively) after landing compared with their levels before flight.[89] Catecholamines, most consistently epinephrine, which were evaluated after the first five Apollo missions, were also higher after, than before, flight.[86] The aldosterone level in the urine of ASTP crewmembers was greater and the antidiuretic hormone and norepinephrine levels were lower after landing than before flight; urine levels of Na^+, K^+, and Cl^- were reduced after this flight, as after other flights. The Soviets also reported that urine levels of Na^+ and Cl^- were lower after short flights than before, and urine levels of aldosterone and catecholamines (especially epinephrine) were elevated after flight.[90]

During the long-duration Skylab flights, levels of Na^+, K^+, Cl^-, aldosterone, and cortisol in the urine were higher than these levels before flight; and urine levels of antidiuretic hormone, epinephrine, and norepinephrine tended to be lower than they were before flight. Antidiuretic hormone levels in the urine collected from two cosmonauts on flight days 216 to 219 were higher than before flight.[95] Immediately after landing, the Skylab crewmembers' urine levels of aldosterone, epinephrine, norepinephrine, and cortisol were elevated above preflight levels; and Na^+, K^+, Cl^-, and antidiuretic hormone were lower than preflight levels.[91] The Soviets also reported that urine levels of aldosterone,[99] epinephrine, and norepinephrine were elevated after long-duration space flight, and the elevation in catecholamines (except epinephrine) was sustained longer after a long-duration space flight than after short flights.[98]

The most recent U.S. data are from the NSTS program. Blood and urine samples were obtained from crewmembers before and after NSTS flights 1 through 4. However, as Leach[96] points out, these data must be interpreted with caution for at least two reasons: (1) as with previous space flights, the preflight blood samples were drawn in early morning when the astronauts were fasting, but the first postflight samples were drawn later in the day after the astronauts had eaten; and (2) most crewmembers consumed large quantities of fluid before the postflight measurements. Nevertheless, after flight, urine Na^+ and Cl^- levels were significantly decreased and urine aldosterone was significantly increased, as compared with before flight. Urine levels of antidiuretic hormone, catecholamines, and cortisol were unchanged after flight compared with those levels before flight. After a new urine collection device had been developed for use in flight, one NSTS crewmember collected all urine for 4 d before flight, 6 d during flight, and 3 d after return.[97] Antidiuretic hormone excretion increased sevenfold in the urine sample collected from 3 h 22 min before launch to 1 h 50 min after launch, but returned to preflight levels thereafter. The time of the early transient rise would include responses to the cephalad fluid redistribution resulting from the prelaunch posture. Seventeen hours later, during flight, the excretion of cortisol showed a large (13-fold) but transient increase, accompanied by a transient increase in K^+ excretion. Then, on the last day of flight, aldosterone levels in the urine were tripled. Immediately after landing, urine volume, aldosterone, K^+, and Cl^- were sharply decreased. During the first days of flight, this subject experienced space motion sickness, which probably affected some values, particularly antidiuretic hormone.

ii. Plasma

Blood samples drawn immediately after three Gemini flights (of 4, 8, and 14 d) had higher levels of K^+, aldosterone, and catecholamines than did blood samples drawn before the flights.[86] After the Apollo missions, plasma K^+ was decreased and angiotensin I was elevated.[89] After the first five Apollo flights, total body potassium, which was measured by a $^{40}K^+$ technique, was reduced by 20%, as compared with that before flight.[86] Angiotensin I and aldosterone measured in the plasma of three ASTP crewmen after landing were greater than those levels in blood drawn before flight, which was consistent with data from other flights. Plasma aldosterone of the crewmembers of NSTS flights 1 through 4 was increased significantly and their K^+ and Na^+ decreased significantly after flight compared with levels before flight.[96] Plasma angiotensin II was unchanged. Aldosterone levels remained elevated and K^+ was depressed several days later. Based on blood samples drawn from 20 cosmonauts before and after their short-duration flights, Grigoriev et al.[90] reported that blood cortisol was greater after flight than before, and blood prostaglandins A and E (which they call the depressor prostaglandins) were reduced after flight. Davydova et al.[98] reported that catecholamine (especially norepinephrine) levels in the plasma were elevated after flight. They attribute this to reduced deactivation, — not increased synthesis.

Blood samples have been drawn during long-duration flights, as well as before and after. Plasma K^+, cortisol, and angiotensin I (usually) were elevated above preflight levels during and immediately after the Skylab flights; and plasma Na^+, Cl^-, and ACTH were reduced below preflight levels during and immediately after flight. Plasma levels of aldosterone were depressed during flight but increased above preflight levels immediately after landing.[91] Tigranian et al.,[99] having found no correlation between flight duration and hormonal changes, combined the data from ten cosmonauts on Soviet flights of 73 to 185 d. They report that plasma renin activity and plasma aldosterone were depressed after these long flights (although these levels were increased after short flights, and the Skylab data showed an in-flight depression but a postflight elevation in plasma aldosterone). Plasma catecholamines were increased after these long-duration flights, as after short-duration flights, but after the long flights epinephrine was elevated more than norepinephrine.[98] Blood drawn from two cosmonauts on day 217 of their 237-d mission showed plasma renin activity higher than before flight.[95] According to these investigators, both cosmonauts were in positive fluid balance and positive Na^+ balance with activation of hormonal systems that regulate fluid and electrolyte homeostasis.

2. Simulated Weightlessness
a. Studies Less than 24 Hours
i. Horizontal and Head-Down-Tilt Bedrest

Reversals of the initial cardiovascular responses to bedrest appear to be explained by responses of other body systems to the initial fluid overload in the central circulation. Urine output, especially for the first 6 h, [32,34-36] and sodium excretion[67] are increased. After 24 h, total body water is reduced by 1.3 l. Total blood volume is decreased[36] by about 500 ml after 6 h at 5° head-down tilt.[32,34] Volicer et al.,[67] using the Evans blue dye-dilution technique, determined that over 300 ml of fluid are lost from the plasma in the first 24 h, which is consistent with the increased hematocrit.[32] Hargens et al.,[35] using a wick-catheter technique, determined that interstitial fluid pressure in subjects' tibialis anterior muscles was reduced after subjects were tilted 5° head down for 8 h, indicating that part of the fluid that shifted to the upper body came from the interstitial spaces. Neither blood osmotic pressure nor tissue colloid osmotic pressure was changed in that experiment.

Thus, the body's central fluid status appears to be regulated to values that are normal for the upright posture on Earth. This may be mediated by the Henry-Gauer reflex and by the secretion of atrial natriuretic factor during the volume overload of the central circulation that occurs immediately after the initiation of head-down tilt. This hypothesis has been tested by monitoring hormone levels in the plasma, but results from several studies have been conflicting. Nixon et al.[34] and Blomqvist et al.[36] observed trends toward reduced levels of plasma renin activity, aldosterone, and antidiuretic hormone after their subjects had been tilted head down for 6 h. These levels returned to control values by 24 h. However, Volicer et al.,[67] whose subjects were also tilted 5° head down, observed no change in the subjects' plasma renin activity or aldosterone after 6 h of head-down tilt, and an increase in these levels at 24 h. Dallman et al.[100] attempted to reconcile these differences in a study of eight men tilted 6° head down for 7 d. On the first day, plasma levels of arginine vasopressin, renin activity, and aldosterone of their subjects were depressed; by 24 h, plasma renin activity had increased back to "control", while aldosterone levels were still low. However, their subjects were standing during the control testing, which may account for some differences between the results of this study and those of the others. The question of how hormones respond to the central fluid overload caused by head-down tilt should be resolved, and atrial natriuretic factor should be monitored in future studies to determine its role.

ii. Water Immersion

Urine flow and electrolyte excretion also increase when subjects are immersed in water. They were proportional to the depth of the immersion[43] and were greater when subjects' blood volume had been expanded by infusion than when it had been contracted by blood withdrawal.[101] Epstein et al.[40] made certain that their subjects were hydrated before immersion and that they remained hydrated by having them ingest 200 ml of water each hour during immersion. These subjects lost 2.2 kg after 8 h of immersion, despite the fluid intake. Their urine flow increased by the end of the first hour, and their urine Na^+ concentration increased by hour 2, while urine K^+ increased only transiently. Serum osmolality, Na^+, and K^+, however, did not change. Urine flow during water immersion was twice as great during the day as at night.[102,103] In addition, although increases in urine flow, Na^+ excretion, and osmotic clearance were greater with water immersion during the day than at night, creatinine clearance (indicating glomerular filtration rate) did not differ between control and immersion or between night and day.[104] Thus, it was concluded that water-immersion-induced diuresis results entirely from osmotic causes.

Plasma volume, blood volume, extracellular volume, and interstitial fluid were all reduced with water immersion, and hemoglobin and hematocrit were increased. In fact, hemoglobin and hematocrit were higher as a consequence of water immersion than as a consequence of head-down tilt.[69] When variables were measured during immersion to the umbilicus, chest, and neck, plasma volume decreased only when immersion was to the chest or neck, whereas hemoglobin and hematocrit increased in all immersion experiments.[43] Greenleaf et al.[105] measured plasma volume (T-1824) and extracellular fluid (bromide space) of four subjects before and after 8 h of water immersion. Although their subjects drank 200 ml water an hour, their water balance was a negative 1234 ml after immersion. Extracellular volume decreased by 2230 ml, which was 19% plasma volume and 81% interstitial fluid. Blood volume decreased by 550 ml (10%). Of this loss, 78% was plasma and 22% (119 ml) was red cell volume. The investigators concluded that, since extracellular volume was reduced by more than the loss of body fluids and red cell volume, fluid must have shifted from the extracellular compartment to the intracellular compartment. On the other hand, Khosla et al.[106] measured interstitial fluid pressure in an antecubital fossa using the wick method and observed that interstitial fluid pressure decreased by 2.07 cm H_2O. They attributed this to a shift of interstitial fluid, supplied partly from the cells, into the blood because of a change in the transcapillary pressure gradient in the submerged body, — and not to a loss of fluid in urine. They disagree with the conclusion of Greenleaf et al. that extracellular fluid entered the cells. However, they did not measure changes in fluid compartment volumes.

Hormonal responses to water immersion were greater than those to horizontal or head-down-tilt bedrest. Epstein[107] reported that when subjects were immersed in water, their plasma renin activity was reduced to 38% of the preimmersion values, plasma arginine vasopressin was suppressed (with no change in plasma osmolality), and renal prostaglandin E was excreted. From the study of responses to different levels of water immersion, Norsk et al.[43] reported that plasma arginine vasopressin levels were inversely correlated with central venous pressure. In another study,[101] these same investigators observed that plasma arginine vasopressin levels were more highly correlated with systolic pressure and mean arterial pressure than with central venous pressure; they concluded that the arterial baroreceptors are more important than the cardiopulmonary mechanoreceptors for the regulation of arginine vasopressin in humans. Gerbes et al.,[108] using 16 men and 9 women who were immersed for 60 min, observed that plasma atrial natriuretic factor increased from 6.0 ± 0.6 fmol/ml to 13.6 ± 2.6 fmol/ml, while plasma cyclic GMP was also increasing and plasma renin activity and aldosterone were suppressed.

Endocrine responses to water immersion do not vary diurnally as do urine responses. In a study of the responses of six subjects to immersion in water, plasma renin activity was reduced during water immersion both day and night (control levels were greater in the day), and atrial natriuretic factor increased within 20 min after water immersion both day and night (control levels day and night did not differ).[102] In another study,[104] reductions in aldosterone and antidiuretic hormone with water immersion were the same during the day as at night.

Thus, the diuresis during water immersion appears to result at least in part from reduced activity of plasma renin, aldosterone, and arginine vasopressin, and from increased activity of atrial natriuretic factor.[43,101,102,104,105,107,108] The suppression of arginine vasopressin during water immersion may not occur when subjects are hydrated before immersion.[108a] Other influences, such as sympathetic nervous system outflow to the renal vessels or prostaglandin activity, may also be involved, but the data are conflicting. Mano et al.[109] reported that when their subjects were immersed the amplitude of sympathetic nervous system activity measured directly in the soleus was reduced; and Bonde-Petersen et al.[69] reported that catecholamine levels of their four subjects were lower during water immersion than during head-down tilt. On the other hand, Epstein et al.[110] reported that neither epinephrine nor norepinephrine levels differed from control levels after 4 h of immersion to the neck; however, they used only four subjects each for seated control and immersion.

b. Studies Longer than 24 Hours
i. Horizontal and Head-Down-Tilt Bedrest

In agreement with data from weightlessness, most of the data from bedrested subjects indicate that their plasma volume and total body water were reduced after several days.[70,73,82,111-113] Hyatt and West[43a] studied eight male subjects who performed both horizontal bedrest and 5° head-down bedrest. Plasma volumes of these subjects were reduced after both bedrest periods, with no difference between the magnitude of the reduction after horizontal bedrest and that after head-down bedrest. Plasma volumes of eight men were reduced more than 10% after 2 weeks of horizontal bedrest.[111] Plasma volumes of five men were reduced 8% after 20 d of horizontal bedrest[70] and were reduced 11% after they were tilted 6° head down for 8 d. Plasma volumes of these five men were reduced further after 30 d.[79] On the other hand, after 1, 2, and 3 weeks, plasma volumes of six men studied by Chobanian et al.[73] were reduced 305, 261, and 175 ml, respectively, and the decrease in plasma volume that Hyatt et al.[114] observed after 1 week of bedrest was no longer significant after 2 weeks.

As reported above, space flight crewmembers' plasma volumes were measured before and after several of the Gemini and Apollo flights and the three Skylab flights. In addition, plasma volumes were estimated from hematocrit and hemoglobin values before and after Apollo, Skylab, and NSTS flights.[115] In all cases, plasma volumes were smaller after, than before, space flight. However, we have observed from these data that plasma volumes tend to be reduced more after flights of 4 to 8 d than after longer flights of 10 to 14 d.

Analysis of fluid compartments is difficult to perform during space flight, because of the invasiveness of the methods and analytical requirements. Assuming that bedrest is a good simulation of space flight, these data can provide information about in-flight status. As such, it is noteworthy that two of the reported studies of progressive changes in plasma volume during bedrest demonstrate a recovery of volume during the second and third weeks of bedrest, after an initial decrease during the first week, which agrees with our observations based on pre- and postflight measurements from flights with different durations.

Johnson[116] compared changes in blood volume, plasma volume, and red cell mass after space flight with changes in these variables after bedrest periods of comparable durations. Nine astronauts who flew on three Apollo missions lost more blood volume than did six men who spent 14 d in bed for a study by Johnson and Hoffler (415 ± 60 ml vs. 310 ± 72 ml); whereas the bedrested men lost slightly more plasma volume than the Apollo astronauts (197 ± 74 ml vs. 158 ± 68 ml), indicating the astronauts lost more red cell mass. Blood volume loss of the three astronauts on the 28-d Skylab mission was approximately equal to that of six men who spent 28 d in bedrest for a study by Hyatt and West[117] (619 ± 115 ml vs. 632 ± 69 ml); but, after 28 d, the men in the bedrest study had lost more plasma volume than the astronauts had (488 ± 73 ml vs. 307 ± 121 ml), indicating that these astronauts also lost more red cell mass than bedrested men did. Johnson[116] stated that plasma volume varies diurnally, being lower early in the morning than later in the day. Thus, he explained, since all measurements — except postlanding measurements — were made in the early morning, the true decrease in plasma volume after space flight might have been relatively greater than revealed in these results.

Several interventions have lessened the reduction in plasma volume that occurs during bedrest, and these interventions may be appropriate for space flight if it is deemed desirable to lessen the plasma volume loss that occurs with space flight. Dynamic exercise is one such intervention. Both Stremel et al.[78] and Greenleaf et al.[79] reported that if subjects performed regular dynamic exercise for 30 min twice a day during a period of bedrest, their loss of plasma volume was significantly reduced. Neither static exercise nor isokinetic exercise performed for 30 min twice a day was effective in reducing plasma volume loss.[78,79]

LBNP is another possible intervention to reduce the loss of plasma volume. Stevens et al.[82] found that LBNP restored the plasma volume of subjects who underwent 10 h/d of −30 mmHg LBNP for several days at the end of a 28-d bedrest period; subjects who did not have LBNP had a plasma-volume loss.

Drugs also may prevent the loss of plasma volume. Ten control subjects studied by Hyatt et al.[114] lost plasma volume during 14 d of bedrest, but ten subjects who received 9-alpha-fluorohydrocortisone did not lose plasma volume, despite significant diuresis and saluresis. The investigators concluded that the subjects lost fluid from their extravascular compartments. Stevens et al.[112] also observed that administration of 9-alpha-fluorohydrocortisone restored plasma volume of their subjects. In addition, when exogenous estrogen was administered to 12 women who were horizontal for 11 to 12 d, their plasma volume did not change; however, their plasma volume decreased by 20% when no estrogen was administered.[118,119]

Finally, use of mechanical devices may reduce the loss of plasma volume. Occlusive cuffs around the thighs may be one such intervention.[112] Another potential countermeasure to fluid loss during space flight or simulated space flight is a "reverse gradient garment", which induced pooling of blood in the veins of the extremities in quantities similar to the pooling that occurs with standing.[81,120] Four men who wore this garment during a period of bedrest experienced less plasma-volume reduction after bedrest than did four control subjects who did not wear the garment.

These studies of techniques for potentially reversing the plasma-volume losses during simulated space flight provide guidance for developing countermeasures to be used with actual space flight and provide information the clinician may use in helping patients to reambulate.

As with studies of electrolyte and hormone levels during short-term bedrest, some differences exist among studies of the effects of long-term bedrest on electrolyte and hormone levels; and some differences exist between data from actual space flight and those from simulated weightlessness. After 6 d of 6° head-down tilt, the subjects of Dallman et al.[100] had no change in their plasma levels of Na$^+$, K$^+$, arginine vasopressin, or ACTH. Their plasma renin activity was elevated, and their aldosterone was depressed. Thus, these investigators observed a change in the coupling of these two hormones. Because plasma renin activity and aldosterone responded normally to saline loading and to ACTH infusion, they concluded the change in the plasma renin activity-aldosterone relationship resulted from "unidentified factors" that may affect angiotensin II. More recently, Annat et al.[44] reported that plasma renin activity and aldosterone levels of their subjects who were tilted 6° head down for 4 d, were increased. Thus, no apparent change occurred in the coupling of the hormones. Their subjects were supine for control tests, however, and Dallman's were standing. Annat et al.[44] also report that after their subjects were tilted for 4 d, urine flow and the urine Na$^+$ to K$^+$ ratio were decreased, while plasma electrolytes, osmolality, creatine, and hematocrit were unchanged compared with values on the first day. These investigators observed a circadian rhythm in plasma renin activity and plasma aldosterone during the 4 d of head-down tilt, an observation that differs from the results of the short-duration water-immersion study of Miki et al.[102]

ii. Water Immersion

After subjects had been immersed in water for 7 d, their plasma renin activity was increased. Sodium loss occurred on the first day, and K$^+$ loss started on day 5 when the aldosterone level in the urine increased.

Greenleaf[121] hypothesized that the headward fluid shift induced by weightlessness increases central venous pressure enough to stimulate the release of atrial natriuretic factor, but not enough to stimulate the release of vasopressin. This would cause a natriuresis and a small increase in urine flow. The plasma would become more dilute, and fluid would move to the intracellular compartment, causing edema and inhibiting thirst. Thus total body water would decrease slowly until a new steady state was reached. Since head-down tilt induces diuresis more gradually than water immersion does, it may be a better model for fluid and electrolyte changes during space flight.

D. BARORECEPTOR REFLEX RESPONSIVENESS

It has long been suspected that reflex control of the cardiovascular system is changed as a result of exposure to weightlessness or to other characteristics of the space-flight environment. Carotid sinus baroreceptor reflex function has been studied after simulated weightlessness, and several studies have been initiated to study the function of the carotid sinus baroreceptor reflex during or before and after space flight.

1. Space Flight

A study of the influence of space flight on the function of the carotid sinus baroreceptor reflex was initiated with STS flight 27 in December 1988.[122] In this study, the carotid-sinus baroreceptors of crewmembers were stimulated using the neck-cuff method before and after flight, and the R-R interval from the ECG was measured on a beat-by-beat basis and related to the carotid sinus transmural pressure at each beat to evaluate the responsiveness of the reflex. To date, eight astronauts have been studied. Preliminary results suggest that carotid-sinus baroreceptor reflex function is impaired after as few as 4 d of orbital flight. These changes may affect orthostatic function. An in-flight study of baroreceptor reflex function is planned for the Space Life Sciences mission (SLS-1) in 1990.

2. Simulated Weightlessness

Convertino et al.[123] tested the cardiac response of 11 men to stimulation of their carotid sinus baroreceptors by the neck-cuff technique before, during, and after these men spent 30 d at 6° head-down tilt. Both the maximum slope of the relationship between carotid sinus pressure and

R-R interval of the ECG and the range of pressures in which the carotid sinus baroreceptor reflex responded decreased progressively throughout bedrest. Neither arterial pressures nor catecholamine levels changed with bedrest. Plasma volume decreased 15% by day 3 of bedrest and did not change further. The change in plasma volume was not significantly related to the change in slope of the carotid sinus pressure/R-R interval curve.

These data suggest that the function of the carotid sinus baroreceptor reflex is changed with both actual and simulated weightlessness. This change may result from influences other than the reduced plasma volume, such as a reduction in the range of arterial blood pressures to which the baroreceptors are exposed during space flight and bedrest.

E. ORTHOSTATIC FUNCTION

1. Space Flight

In this section the evidence of altered orthostatic function during space flight and at the earliest medical evaluations after flight is summarized. Standardized provocative tests of orthostatic function have been performed before and after all U.S. flights since the Mercury MA-9 mission. Test methods have included passive standing, head-up tilt, and LBNP. None of these tests has been performed immediately after landing; a readaptation period of several minutes to several hours, during which crewmembers have stood and walked, has always preceded testing. Evaluation of orthostatic function in flight has been limited to LBNP tests during space-station missions.

Postflight orthostatic dysfunction in U.S. astronauts was first suspected after the 9-h Mercury mission and then confirmed after the 34-h Mercury mission.[48] After the 9-h mission, the astronaut egressed from the spacecraft to the carrier deck unaided with no gait disturbance, lightheadedness, or dizziness.[14] However, 1 h later, his heart rate increased from 70 to 100 bpm as he rose from the supine to the upright position. Concurrently, his systolic pressure fell substantially, although diastolic pressure was maintained. His heart rate and systolic pressure recovered as he lay down again. No preflight lying-to-sitting blood pressures were available for comparison. After the 34-h Mercury flight, the astronaut's heart rate of 132 bpm while lying in the spacecraft may have been related to taking dextroamphetamine in flight, but his rate increased by 42% (to 188 bpm) when he stood on the deck of the recovery ship.[51] Blood pressures when he was in the spacecraft were 101/65 mm Hg and 105/87 mm Hg, but the automatic blood pressure device detected no pressure pulses when he stood on the ship. After about 1 min of standing, the astronaut became pale and diaphoretic, swayed slightly, and reported lightheadedness, dimming of vision, and tingling of the feet and legs. The average increase in his heart rate during preflight tilt tests was 29%.[51] Systolic blood pressure increased only 3%, diastolic pressure increased 18%, and pulse pressure narrowed by 38%. During the earliest postflight tilt tests, the astronaut had no presyncopal symptoms, but his heart rates throughout each test exceeded the preflight averages. The average increase in heart rate during these postflight tilt tests was 48%. Blood pressure responses were similar to those in the preflight period.

Although no crewmembers during the Gemini program reported presyncopal symptoms or syncope upon egress,[4] at least one fainted during passive tilt after his flight,[156a] and the postflight passive tilt tests consistently evoked greater increases in heart rate (17 to 105%) and a greater narrowing of pulse pressure than did preflight tests.[4]

During the Apollo program, orthostatic function was evaluated by graded LBNP tests and passive stand tests.[18,19] The protocol consisted of three LBNP stages: –30 mm Hg (5 min), –40 mm Hg (5 min), and –50 mm Hg (5 min). Most crewmembers had three preflight orthostatic tests, and the first postflight tests were performed 2 to 7 h after splashdown. At 50 mm Hg LBNP stress, Apollo astronauts had an average increase in heart rate of 24% before flight, along with a 10% fall in systolic pressure, a 6% increase in diastolic pressure, and a 31% narrowing of the pulse pressure. (These percentages were computed from the raw data of the NASA Cardiovascular Laboratory.) After flight, heart rate increased 57% and pulse pressure narrowed by 45%, mostly from an 18% fall in systolic pressure.[18] One of the 18 crewmembers tested became presyncopal before the 50 mm Hg negative pressure was reached; five additional crewmembers developed presyncopal symptoms at −50 mm Hg. The Apollo passive-stand protocol consisted of 5 min of supine rest followed by 5 min of passive standing, during which crewmembers leaned against a wall in a relaxed manner with their heels 6 in. from the wall. The increased heart rate response after flight equalled that noted during the LBNP tests, but narrowing of the pulse pressure with the stand test resulted mostly from an increase in diastolic pressure. From combined LBNP and stand-test data, Hoffler and Johnson[19] found a statistically significant correlation of 0.52 between the percent change in resting heart rate after flight and the percent change in orthostatically stressed heart rate after flight.

Charles and Bungo[124] observed that the heart rate increase during postflight orthostatic tests peaked after flights of 8 d and was less after flights of 12 to 14 d. The ingestion of salt tablets and water before landing simply shifted the heart-rate curve downward. Furthermore, the ratio of the dimensions of the heart to the dimension of the thorax measured from an X-ray photograph (cardiothoracic ratio) was maximally reduced after flights of 3 d but did not differ from preflight values after flights of 12 to 14 d. They concluded that (1) the loss of orthostatic tolerance and "heart size" are multiphasic processes, with physiological mechanisms that vary with time; and (2) the loss of body fluid volume is only one component of the development of post-space flight orthostatic intolerance.

Soviet experience with postflight orthostatic tolerance is similar to the U.S. experience. Heart rates of the cosmonauts during reentry and landing on the Vostok/Voskhod missions exceeded their rates during preflight centrifugation tests that simulated the gravitational stress of landing. Heart rates averaged 10 bpm higher after the two 1-d missions, 30 to 32 bpm higher after flights of 3 to 4 d, and 62 bpm higher after the 5-d flight.[125] The crewmembers of these missions responded to a "passive orthostatic test with measured stress" that was performed after landing with greater heart-rate increases and longer recovery times than they had before flights,[49,125] but they successfully completed the tests.[126] The passive orthostatic test consisted of 10 min at 88° head-up tilt. The most definitive test of orthostasis, however, may have been the act of standing. Two of the cosmonauts experienced 26 to 47% reductions in stroke volume and cardiac output, and decreases of 10 to 24 mmHg in pulse pressure when they stood.[126] Cosmonauts of six Soyuz flights that lasted 3 to 5 d apparently experienced some postflight changes in orthostatic responses during "active and passive" transition from the supine to the upright position, but the most severe responses were noted in crewmembers of the 18-d Soyuz 9 flight.[10] Both cosmonauts had to be carried from the space craft after flight.[1] They had difficulty "maintaining equilibrium" for about 3 h after landing,[127] that is, they experienced dizziness, weakness, and tachycardia when they tried to stand and preferred to lie down.[10] During "active" orthostatic tests, which consisted of heart rate and blood pressure measurements in the lying, sitting, and standing positions $1^1/_2$ to 2 h after landing, both cosmonauts responded with blanching of the face.[128] Their heart rates increased by 33 and 22%, and pulse pressures decreased by 50 and 43% after 5 min of sitting; and after 5 min of standing, heart rates increased by 63 and 65%. Korotkoff sounds were not detectable in one of the cosmonauts when he stood and were audible only once in the other, at which time the pulse pressure was reduced by 71%. One of these cosmonauts had flown a previous 4-d mission with no postflight loss of orthostatic tolerance. The experience of the Soyuz 9 crew was anomalous, in that subsequent longer Soyuz missions were not characterized by such pronounced intolerance.[1] Both crewmembers had exercised regularly in flight.[10]

For all NSTS flights through 1988, orthostatic function was assessed by preflight and postflight passive-stand tests during which the heart rate was monitored continuously and the blood pressure was measured at 1-min intervals. Fluid was sometimes consumed before the postflight tests, which have been performed several hours after egress from the orbiter. Bungo and Johnson[54] reported data from the eight crewmembers of NSTS flights 1 to 4, which lasted

2.3 to 8 d. From their published data, we have computed that passive standing before flight produced an average 22.1% increase in heart rate, 3.0% decrease in systolic pressure, 1.5% increase in diastolic pressure, and 11.0% decrease in pulse pressure; whereas passive standing after flight produced a 45.2% increase in heart rate, 8.8% decrease in systolic pressure, 3.5% decrease in diastolic pressure, and 18.2% decrease in pulse pressure. One crewmember developed presyncopal symptoms after flight.

Forces in the head-to-toe direction of greater than 1 G are experienced by NSTS crewmembers during reentry and landing. Although no presyncopal symptoms or fainting have been reported to interfere with landing the orbiter, data compiled from the 26 missions completed through 1988 show eight occurrences of presyncopal symptoms or actual syncope prior to egress and seven occurrences of presyncope or syncope during postflight orthostatic testing. Altogether, one or more crewmembers were presyncopal or syncopal during either egress or postflight orthostatic tests after 35% of the 26 missions. With a few exceptions, the evidence for orthostatic intolerance after space flight has been most convincing after the NSTS flights. We suggest several possible causes: this is the only vehicle in which crewmembers are exposed to forces of greater than 1 G in the head-to-foot direction during landing; the incidence of space motion sickness during flight has been greater with NSTS than it was with the earlier, more confining vehicles; the incidence of both space motion sickness and postflight intolerance to orthostatic stress has increased since the crewmembers started wearing the new launch and landing suit (STS-26), which tends to be hot and to require considerable head motion in donning and doffing.

Responses to orthostatic stress were monitored during the long-duration space station flights as well as before and after these flights. A total of 138 LBNP tests were performed in flight on the three Skylab missions.[56] The protocol was derived from that used before and after Apollo flights and consisted of five LBNP stages: –8 mm Hg (1 min), –16 mm Hg (1 min), –30 mm Hg (3 min), –40 mm Hg (5 min), and –50 mm Hg (5 min). During all three missions, the initial in-flight LBNP tests, which were usually performed on the fourth or fifth day of flight, showed greater heart rate responses than had been observed before flight. During the longer missions, the heart rate response to LBNP decreased some after about 6 weeks, but not to preflight levels. The calf volume of crewmembers on the 84-d mission increased more than twice as much during in-flight LBNP as it had before flight, especially at low levels of LBNP, and was slow to return to pre-LBNP levels. This was probably because the leg veins were relatively empty and flattened before LBNP, so a very low pressure could induce filling until the veins reached a circular configuration. Furthermore, crewmembers reported that LBNP was subjectively very stressful throughout the flight;[56] the greater transfer of blood to the legs from a total blood volume that was reduced below normal levels for the 1-G environment may have compromised the central circulation. In-flight responses to LBNP were good predictors of postflight responses, and heart rates during the first postflight LBNP test approximated those during the last in-flight test.[56]

Postflight orthostatic intolerance was no more severe after the Skylab missions than it was after shorter flights, but a longer time was required for orthostatic responses to return to preflight levels.[56] In fact, the most pronounced responses to LBNP after Skylab were often on later days after landing, not on the landing day itself.

LBNP, using a "trouser-shaped" vacuum bag (named "Chibis"), has been performed during all missions to the Salyut and Mir space stations. The Soviet LBNP test uses a protocol of 2 min at –25 mm Hg and 3 min at –35 mm Hg.[129] On 30-d and 63-d missions to Salyut 4, tolerance tests were performed once a week, and LBNP was performed for 30 min a day on the last 4 d, with two 30-min sessions plus ingestion of salt and water on the last day.[17] Increases of the cosmonauts' heart rates in response to LBNP were greater during flight than they had been before flight, except when the LBNP was accompanied by water and salt ingestion. Yegorov et al.[52] report that cosmonauts' responses to LBNP were more pronounced during all the long-duration missions from 65 to 237 d than they were before flight: heart rate and peripheral resistance increased more; and left-ventricular ejection time, vascular tone and muscle tone, and tissue and venous pressures decreased more. Gazenko et al.[129] concluded that this greater reactivity to provocative tests during flight than before flight resulted from the cardiovascular conditions before the test. They also reported that cardiovascular reactivity to provocative tests did not increase with time, which they interpreted as indicating that the cardiovascular systems of the crewmembers adapted to microgravity. Bogomolov et al.[23] reported that in-flight responses to LBNP of the cosmonaut who spent 326 d in space were similar to his responses before flight, and they concluded that the stability of his cardiovascular system remained adequate for 326 d. In agreement with U.S. Skylab data,[56] testing of crews of the 49-d and 17-d missions to Salyut 5 indicated that individual responses to in-flight LBNP tests were predictive of postflight responses to orthostatic stress.[130] Several crewmembers had no problems with LBNP during flight, and they experienced no postflight orthostatic intolerance; one crewmember had decreased endurance to LBNP starting on day 23, had two in-flight tests stopped by physicians, and experienced decreased orthostatic stability after flight.

As with U.S. data, the Soviet data indicate that, although orthostatic tolerance is less after flight than before, it is no worse after long-duration flights than it is after shorter flights. In fact, Bogomolov et al.[23] report that instability of the cosmonaut who spent 326 d in space was no more severe than that observed after previous shorter flights.

On January 11, 1989, the Soviet News Agency TASS reported that the two cosmonauts who had spent a full year in space experienced minor transient vestibular problems immediately after touchdown. The report indicated that both cosmonauts walked 3 to 4 km and swam 400 to 500 m starting the fifth day after landing. Muscle atrophy and some calcium loss were reported, but provocative testing for tolerance to orthostatic stress was not mentioned. In contrast to the TASS report, *Forbes* (December 1988) recently cited Oleg At'kov, a physician and former cosmonaut, as saying that the Soviets' extensive experience with long-duration space flight has shown that humans cannot stay weightless in space more than 12 months without risking permanent damage to normal cardiovascular function, despite the use of vacuum trousers designed to pool body fluids from the upper chest and head to the lower body.

2. Simulated Weightlessness
a. Studies Less than 24 Hours
i. Horizontal and Head-Down-Tilt-Bedrest

LBNP has been used by several investigators to evaluate responses and tolerance to orthostatic stress after short-duration simulated weightlessness. The following studies all used graduated protocols with a minimum pressure of –50 mm Hg. In one study, eight men responded similarly to LBNP after they were tilted to 6° head down for an hour as they did after a control period in which they were supine without tilt.[131] However, after the head-down tilt, their cardiovascular systems started from a condition of reduced stroke volume, mean stroke ejection rate, and cardiac output; and the further reduction in these variables during LBNP may have compromised their cardiovascular status. In another study in which seven men were tested after 2 h of 6° head-down tilt,[132] four of these tests were terminated early because the subjects developed symptoms of presyncope. However, these subjects were invasively instrumented with right-heart catheterization (see Chapter 1). The subjects were not presyncopal during tests performed several days prior without the invasive instrumentation or the preceding head-down tilt. A third study evaluated subjects after 20 h of head-down tilt. Four of five subjects became presyncopal at –40 mm Hg LBNP, whereas all five had tolerated –50 mm Hg in the control period prior to tilt.[32] Compared with values during LBNP before head-down tilt, heart rates were higher and stroke volumes lower during LBNP after 20 h of head-down tilt in this study[32] and after 24 h of head-down tilt in another study from the same laboratory.

Thus, the changes in responses to orthostatic stress after such brief periods of head-down tilt suggest that a reduction in plasma volume contributes to reduced orthostatic tolerance.

ii. Water Immersion

Short periods of water immersion may decrease tolerance to orthostatic stress even more than bedrest does. Tolerance times of 11 men tested by Hordinsky et al.[133] decreased after 6 h of water immersion but not after 6 h of bedrest. Graveline et al.[103] monitored responses of four men to 12 min of 90° head-up tilt and to acceleration in a centrifuge to $+3G_z$. They compared the subjects' responses to these stresses before water immersion with their responses after 6, 12, and 24 h of immersion. Heart rate increases and pulse pressure decreases during head-up tilt were progressively greater with each post-immersion exposure; after 24 h of immersion, pulse pressure during head-up tilt was only 2 to 6 mmHg. The heart rate response to centrifugation was also greater after immersion, but the time to blackout during centrifugation was not changed after immersion.

b. Studies Longer than 24 Hours
i. Horizontal and Head-Down-Tilt Bedrest

Dietrick et al.[134] observed that healthy subjects were more apt to faint during 65° head-up tilt if they had been bedridden (horizontal) (with pelvis and legs casted) for a week before the head-up tilt than if they had been ambulatory before the tilt. This observation has been confirmed by others, including Chobanian et al.,[73] who saw 3 of 18 subjects become presyncopal during 70° head-up tilt after 3 weeks of horizontal bedrest, whereas none became presyncopal during head-up tilt before bedrest. The greater increases in heart rate[73,114,134,135] and greater decreases in pulse pressure[134] that subjects experienced during orthostatic tests after bedrest, as compared with before bedrest, were similar to the responses before and after space flight. Total peripheral resistance also increased more during orthostatic tests after bedrest than before,[73,114] in partial compensation for the greater decreases in cardiac output (dye dilution), central blood volume,[73,114] and end-diastolic volume.[135]

Muller et al.[136] estimated the volume of subjects' legs with a Whitney strain gauge and measured venous pressure with transcutaneous puncture of a foot vein during LBNP. At the onset of LBNP, both before and during a 7-d head-down-tilt period, venous pressure fell faster than the LBNP pressure, providing a gradient for fluid to move from the tissue spaces to the veins. During steady-state LBNP at –40 mm Hg, venous pressure increased, creating a gradient for fluid to filter from the vessels to the interstitial compartment. This filtration rate, which was greatest during the first days of head-down tilt, combined with the reduced central venous pressure and end-diastolic volume after bedrest, may explain in part the more frequent occurrence of presyncope during orthostatic tests after bedrest than before bedrest.

Convertino et al.[123] observed that subjects who became presyncopal during a post-head-down-tilt stand test had a greater decrease in the slope of the carotid sinus pressure/R-R interval relationship (neck-cuff technique) after 25 d of head-down tilt than did their subjects who did not become presyncopal during the stand test. Their data indicate that losses of orthostatic tolerance after long-duration head-down tilt may result in part from a change in function of the arterial baroreceptor reflex. However, subjects' catecholamine levels were not altered during bedrest, nor were their catecholamine responses to postural change altered. Chobanian et al.[73] also reported that neither norepinephrine levels nor arterial pressure responses of their subjects to norepinephrine infusion were changed after 3 weeks of bedrest. Thus, the decrease in baroreceptor reflex responsiveness is probably not the result of changes in the sympathetic arm of the effector system.

Researchers have studied several countermeasures that are potentially able to reduce the loss of tolerance to orthostatic stress after a period of bedrest. These included exercise, saline, mechanical devices, drugs, electrical stimulation of muscle, LBNP, and combinations of these. Most studies of the effects of exercise training during bedrest have found that neither dynamic ergometry exercise nor isokinetic exercise for up to an hour a day during the bedrest improve postbedrest orthostatic tolerance.[79,80,137] However, one study reports that subjects tolerated a 5-

min stand test better if they exercised during their 30-d bedrest period,[138] and tolerance to centrifugation was improved when isometric exercise was performed during the bedrest period.[137] Smith has suggested that an exercise regimen specifically designed to maintain the function of the skeletal muscle pump might be most effective.[138a] Other countermeasures that have been only minimally effective or ineffective in restoring orthostatic function after bedrest are saline loading administered without other countermeasures;[111,117] mechanical devices, such as cuffs around the thighs[112] or the reverse-gradient garment described above;[120] or 9-alpha-fluorohydrocortisone.[112]

On the other hand, LBNP training during bedrest or LBNP training combined with saline ingestion may effectively restore tolerance to orthostatic stress during and after bedrest. In a study by Hyatt et al.,[117] six men performed two consecutive 1-week bedrests separated by a 2-week recovery period. After the first bedrest period, two of the men received 1000-ml beef bouillon (153.6 mEq Na^+, 1.6 mEq K^+) during 4 h of LBNP "training" at –30 mm Hg, which was performed immediately prior to a graduated LBNP test, and two subjects received just the bouillon before the LBNP test. After the second bedrest period, each subject received the other countermeasure. Plasma volume (measured with ^{125}I RISA) was significantly increased after both countermeasures — twice as much after LBNP plus saline as after saline alone. The combined countermeasure prevented the excessive increase in heart rate usually observed in LBNP tests after bedrest, but bouillon alone did not. In another study, –30 mm Hg LBNP was administered for 10 h each day on several consecutive days after subjects had been in bed for 28 d. These subjects did not experience the increased incidence of presyncope and the elevated heart-rate response during LBNP tests that were experienced by subjects who did not have the LBNP countermeasure.[82]

A combination of countermeasures may be the best protection for retaining orthostatic function after bedrest or space flight. Exercise alone did not preserve the orthostatic tolerance of six subjects who spent 49 d at 4° head-down tilt, and LBNP (four or five sessions of 1 h) induced only minor changes. However, when the subjects received a combination of exercise, LBNP, and saline ingestion, they retained most of their tolerance to orthostatic stress, their strength, and their exercise capacity. The authors concluded that these countermeasures were compatible. In an even longer study,[83] 12 men spent 182 d at 4° head-down bedrest. Six of these subjects received countermeasures of exercise, LBNP, electrical stimulation of muscles, and saline during the bedrest period. The countermeasure subjects experienced less presyncopy during head-up tilt after bedrest did than the six control subjects.[83] However, with so many concurrent interventions, the causes of altered orthostatic function cannot be isolated. According to the authors, these results agree with data from cosmonauts returning from missions of similar durations.[83]

In summary, orthostatic tolerance tested by head-up tilt, LBNP, and the stand test is reduced after horizontal and head-down-tilt bedrest; the duration of tolerance to stress is shorter, heart rate and peripheral resistance increase more, and cardiac volumes are smaller.

ii. Comparison of Horizontal and Head-Down-Tilt Bedrest

Several authors have compared responses to orthostatic stress after a period of horizontal bedrest with those after head-down tilt. Kakurin et al.[139] concluded that head-down tilt from 4 to 12° is a better simulation of space flight than is horizontal bedrest. When in the head-down position, their subjects experienced facial puffiness, feelings of blood rushing to their heads, nasal congestion and impaired breathing, decreased sensations of taste and smell, light vertigo and nausea when they moved their heads suddenly, and some sensory illusions — all of which have been reported by astronauts and cosmonauts. Furthermore, according to these authors, the changes in responses to head-up tilt and ergometer exercise after a 5-d space flight were more similar to those after 5 d of head-down tilt than after 5 d of horizontal bedrest. As a result of these data, most of the recent "hypokinesia" simulations of space flight use head-down-tilt bedrest to simulate spaceflight. On the other hand, Hyatt and West[43a] did not observe differences between their subjects' heart rates during LBNP after 5° head-down tilt and their subjects' heart rates during LBNP after horizontal bedrest ($p < 0.20$). Each of their eight male subjects were bedrested for two 1-week periods with a 2-week recovery period between. As reported above, they also observed no difference in the magnitude of the plasma volume change caused by 5° head-down bedrest from that caused by horizontal bedrest. The initial sensation of fullness in the head during 5° head-down tilt was lost quickly. Leg volume decreased slightly more during 5° head-down tilt ($p < 0.10$). According to Sandler et al.,[140] tolerance to LBNP was reduced after both horizontal and head-down-tilt bedrest, and changes in heart-rate and blood-pressure responses were the same after both simulations. However, end-diastolic volume, stroke volume, and cardiac output (measured with echocardiography) were reduced more during LBNP after 6° head-down tilt than during LBNP after horizontal bedrest.

iii. Water Immersion

Tolerance time to +3 G_z on a centrifuge decreased from 298 s to 130 s (by 56%) after subjects were exposed to "dry immersion" for 28 d.[141] However, their tolerance was maintained when they received countermeasures during the immersion period, including centrifugation at 1 to 2 G_z twice a day for 60 min, exercise twice a day for 60 min, or a combination of these two countermeasures.[141] Shulzhenko et al.[142] further report that tolerance time to +3 G_z centrifugation was decreased more after 7 d of dry water immersion (63%) than after 7 d of 6° head-down tilt (29%).

iv. Mechanisms

The physiological characteristics that theoretically might be primary mechanisms of the postural intolerance after weightlessness or bedrest include blood volume, arterial baroreceptor reflexes, cardiopulmonary receptor reflexes, venous compliance, and the skeletal muscle pump. Evidence that plasma volume, blood volume, and total body water are reduced after real or simulated weightlessness is overwhelming; and this is generally accepted as a contributing cause of the orthostatic dysfunction. Limited data have been acquired regarding arterial baroreceptor reflex function after actual and simulated weightlessness, but these findings suggest strongly that the responsiveness of the reflex is reduced after weightlessness and bedrest, that is, a given change in transmural pressure at the carotid sinuses induces less change in the heart rate. Whether other effectors of this reflex, such as cardiac contractility, peripheral resistance, and venous tone, are also suppressed after weightlessness or bedrest is unknown. Stimulation of cardiopulmonary receptors is certainly affected when an individual is orthostatically stressed, and the dominant role of these reflexes in eliciting vascular resistance to maintain arterial pressure in the critical first moments an individual is in the upright posture, as well as their role in hormonal control of fluid balance, are well recognized. The integrity of these reflexes after exposure to weightlessness has not been carefully examined. The integrity of venous compliance and of the skeletal muscle pump are closely related. Muscle tissue appears to atrophy in weightlessness, which could seriously affect the skeletal muscle pump and, therefore, affect tolerance to orthostatic stress. Little is known about the relationship between muscle atrophy and responses to orthostatic stress, but this cause of orthostatic intolerance would require evaluation by the stand test or its equivalent, possibly including LBNP with a foot support, rather than by LBNP with a saddle support.

Development of effective and efficient countermeasures to post-space-flight or postbedrest orthostatic intolerance requires an understanding of the mechanisms of orthostatic intolerance. The consumption of salt and water before reentry after space flight, which is directed at the decreased plasma volume, has been effective in mitigating post-space-flight orthostatic intolerance to some extent.[57] As the contributions of arterial baroreceptor reflexes, cardiopulmonary receptor reflexes, venous compliance, and the skeletal muscle pump to post-space-flight and postbedrest orthostatic intolerance are better understood, countermeasures can be targeted at these causes.

The rapid advances over the past 10 to 15 years in the development of methods to study cardiovascular function noninvasively have provided sophisticated tools that can be used in space or in the ground-based laboratory to enhance our understanding of the mechanisms of responses and intolerance to orthostatic stress.

IV. SUMMARY AND CONCLUSIONS

In this chapter we have summarized the data of fluid and cardiovascular responses to real and simulated weightlessness that might affect orthostatic function, and we have described the evidence of altered orthostatic function during and after real and simulated weightlessness. Figure 5 is a diagrammatic representation of the fluid and cardiovascular responses to real and simulated weightlessness and the related reflex and hormonal actions. This diagram is based on theory and on empirical observations.

The initial response to the loss of the head-to-foot hydrostatic pressure gradient involves redistribution of body fluids from the legs and lower trunk to the thorax and head until a steady state of pressure and blood flow is reached. During space flight and some simulations of weightlessness, the responses include fullness in the head and a reduction in leg size. This cephalad fluid shifting apparently occurs rapidly and, when the crew is seated in a head-down, feet-up position in the shuttle, may be nearly complete before launch. In fact, central venous pressure measured during flight approaches central venous pressure of subjects standing on Earth, and findings from head-down tilt and horizontal bedrest studies indicate that the fluid shift is completed rapidly: central venous pressure increases transiently and then decreases to control levels by 60 min. The decrease in leg volume starts immediately and is virtually complete after several days in flight.

Cardiovascular function is altered with space flight. Heart rates of crewmembers peak at launch, entry into orbit, and reentry. Average in-flight heart rate values are often higher than those measured before flight, but heart rates of some crewmembers return to, or below, preflight values. Heart rates measured after flight are consistently higher than those measured before flight (from 15 to 60%). Blood pressures during flight are approximately equal to or slightly higher than preflight values; after landing, blood pressures are elevated when crewmembers have ingested salt and water before entry and landing. Echocardiographic measurements indicate that left ventricular and stroke volumes increase transiently on the first day of flight but decrease thereafter during and after flight. Analyses of cardiothoracic ratios from chest X-rays indicate the heart size is smaller (5% for Apollo crewmen) after flight than before flight.

Likewise, cardiovascular function is altered in simulated weightlessness. Immediately after horizontal or head-down bedrest is initiated, heart rate decreases (5 to 20%) and heart volume increases (5%), but shortly thereafter these responses to the cephalad fluid-shift reverse. Responses to short-duration water immersion are more pronounced and do not reverse, possibly because of the continual transmural pressure on the lower body and abdominal veins and the hydrostatic pressure gradient for cerebral venous flow. After a longer period of horizontal or head-down bedrest, the heart rate is sometimes elevated above prebedrest levels, but differences from prebedrest in most of the other cardiovascular variables are small and questionable. Cerebral blood flow may be altered with bedrest, but data to date are not convincing. Changes in the T wave of the ECG correlate with decreases in serum K^+. A striking circulatory response to bedrest is a reduction in the maximal aerobic capacity (28%). Exercise performed during the bedrest period and the combination of exercise performance, LBNP training, saline ingestion, and electrical stimulation of muscle have been effective in reducing or preventing this loss of aerobic capacity.

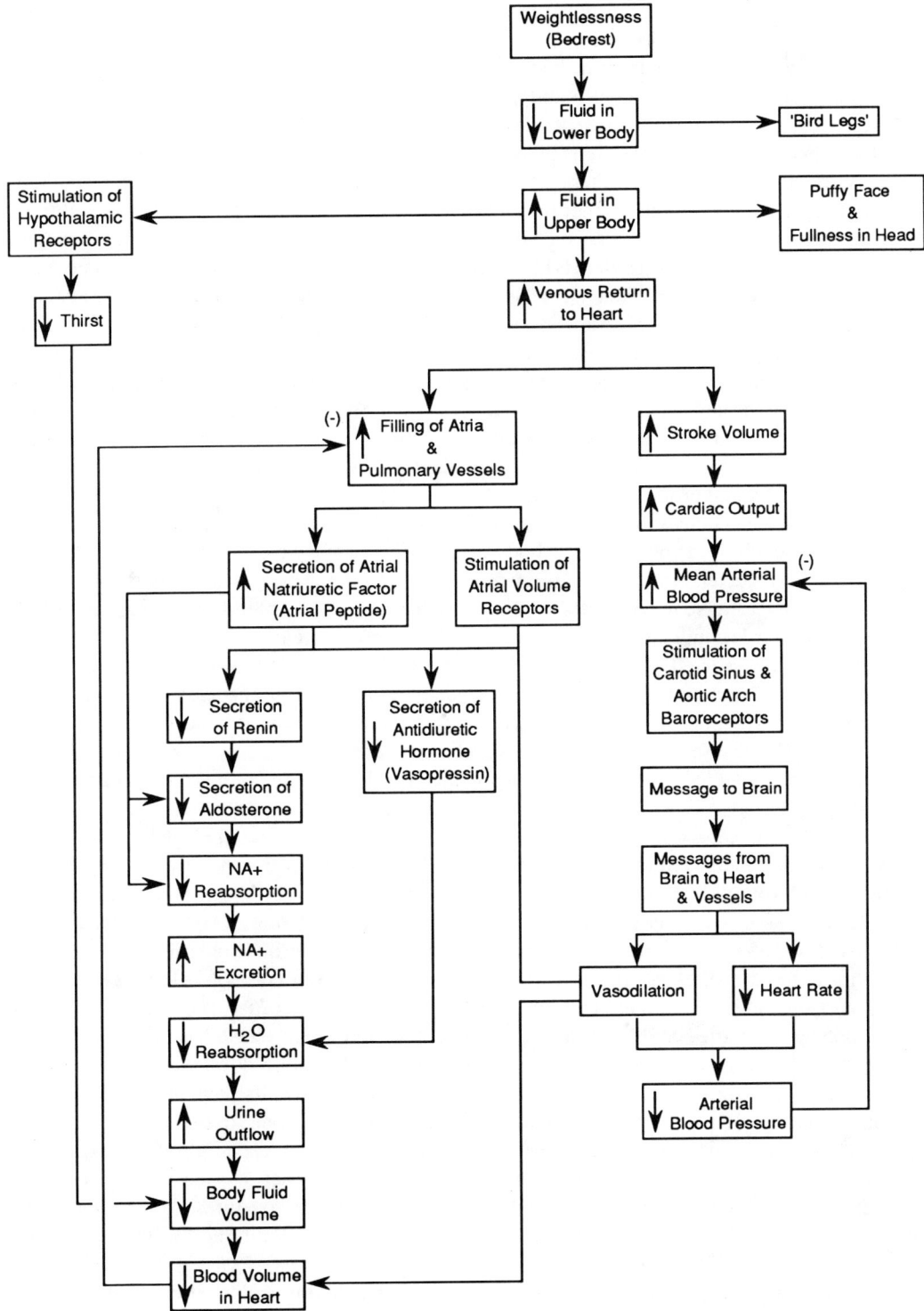

FIGURE 5. Schematic diagram summarizing cardiovascular and fluid responses to real or simulated weightlessness.

While horizontal and head-down-tilt bedrest simulate the fluid shifts of space flight reasonably well, their relationship to in-flight cardiovascular changes is more difficult to evaluate. One complication is the lack of agreement among the simulation studies: in some, heart rate and arterial pressure increase and cardiac output decreases with bedrest; in others, no differences exist from prebedrest values. Many of these differences probably result from differences in experimental design. Investigators have used different tilt angles and different rules governing subject movement and head elevation; different measurement techniques (sometimes invasive, sometimes noninvasive), with some techniques more reliable than others; and different numbers of subjects, such that a response might not be significant in a study with too few subjects. A second problem is in the nature of the space-flight data obtained to date. Most of these data have been recorded before and after flight; particularly after flight, variables such as temperature at the landing site and salt and water consumption almost certainly obscure flight effects. Some data, however, have been obtained during flight, especially during the Skylab missions. A third complication is the time of day when data have been recorded. Many cardiovascular variables conform to diurnal variations. Thus, repeat measurements and measurements to be compared between flight and simulations should be measured at the same time of the crewmembers' or subjects' day; failing that, the time of day should be noted so that data can be corrected and properly analyzed.

Space travellers consistently lose weight during flight (up to 9%). This occurs quickly and represents mostly fluid volume. Total body water (3%), plasma volume (up to 16%), and, after approximately 8 d, red blood cell mass are reduced (11%). Total body water, blood volume, and plasma volume are also reduced after bedrest, but decreases in red blood cell mass after bedrest are small. Some data indicate that plasma volume levels may increase toward control values after several weeks of space flight or bedrest. The decrease in plasma volume with water immersion is greater than that with bedrest and is sustained. Several interventions lessen the reduction in plasma volume that occurs during bedrest, and these interventions — which include dynamic exercise, LBNP and saline ingestion, drugs such as 9-alpha-fluorohydrocortisone and estrogen, and mechanical devices — may provide guidance for developing countermeasures for space flight.

Some differences exist among the results of studies of hormone and electrolyte levels, for both space flight and simulated weightlessness. After some short-duration (up to 14 d) flights, urine aldosterone, antidiuretic hormone, and catecholamines have been elevated; however, NSTS data show no change in antidiuretic hormone or catecholamines. Urine Na^+, K^+, and Cl^- are decreased after short flights. Data obtained during long-duration flights show plasma renin activity is increased, as is urine Na^+, K^+, Cl^-, aldosterone, and cortisol. After long-duration flights, urine aldosterone, epinephrine, norepinephrine, and cortisol are increased above preflight levels; while Na^+, K^+, Cl^-, and antidiuretic hormone are decreased. Both reductions and elevations in plasma aldosterone have been observed after long-duration flights, and this discrepancy may be caused by diurnal variation. Conflicting results have also been reported among investigations of responses to bedrest and head-down tilt. One study showed an increase in plasma renin activity and aldosterone; another showed an increase in plasma renin activity with a decrease in aldosterone. This discrepancy may have been caused by both different measurement times during the day and differences in control postures. We believe these differences should be resolved and the role of atrial natriuretic factor in fluid adjustments to real and simulated weightlessness should be identified.

An understanding of functional changes in the arterial baroreceptor reflex during real and simulated weightlessness is just emerging. The few data that are available indicate that responsiveness of this reflex system is decreased after as few as 4 d of space flight and after short periods of head-down tilt.

Circulatory function in response to orthostatic stress is compromised after even brief periods of space flight and simulated weightlessness. Cardiovascular responses to orthostatic stress have been examined with "stand tests", head-up-tilt tests, and LBNP before and after flight and with LBNP during space-station missions. Responses to LBNP during flight are usually characterized by greater increases in heart rate and leg circumference than before flight. (During LBNP tests on the final Skylab mission, these responses were more than double those during LBNP tests before flight.) Occasionally, especially after they have ingested saline, crewmembers have responded as they did before flight. A crewmember's responses to LBNP during flight are predictive of the magnitude of his response after flight. After flight, crewmembers respond to these tests with double the preflight increases in heart rate and greater decreases in pulse pressure. With a few exceptions, the evidence for orthostatic intolerance after space flight has been most convincing after the NSTS flights. We suggest several possible causes: this is the only vehicle in which crewmembers are exposed to forces of greater than 1 G in the head-to-foot direction during landing; the incidence of space motion sickness during flight has been greater with NSTS than it was with the earlier, more confining vehicles; the incidence of both space motion sickness and postflight intolerance to orthostatic stress has increased since the crewmembers started wearing the new launch and landing suit, which tends to be hot and to require considerable head motion in donning and doffing.

In response to orthostatic stress after horizontal or head-down bedrest or water immersion, heart rate and peripheral resistance increase more (in some studies more than twice as much), and pulse pressure and cardiac index decrease more than they do during orthostatic stress before these simulations of weightlessness. Furthermore, subjects are more prone to become presyncopal or syncopal during orthostatic stress when they have been previously exposed to simulated weightlessness. Neither exercise, ingestion of salty fluids, mechanical devices, nor drugs, when administered singly, are effective countermeasures to the loss of orthostatic function after bedrest. LBNP "training" during bedrest and LBNP plus "rehydration" with saline ingestion, however, protect orthostatic function after simulated weightlessness; and a combination of countermeasures, such as LBNP, saline, and exercise, provide the greatest protection. Controversy still exists regarding whether head-down tilt is a better simulation of actual weightlessness than horizontal bedrest.

Ideally, to understand the physiological effects of weightlessness on humans, reliable data must be obtained during space flight, but difficulties exist in obtaining such data. Especially during and immediately after launch, but also at other times during flight, operational schedules preclude biomedical testing. Also, commercially available instrumentation often is not appropriate for use during flight because it is too large, too heavy, or contains toxic materials. Finally, many of the currently available techniques for obtaining the needed data are invasive or require specially trained personnel. Nevertheless, a concerted effort should be made to obtain reliable data during space flight by developing or obtaining noninvasive monitoring equipment that does not interfere with the other activities of the crew, developing new equipment that can be flown, and sending trained specialists or training astronauts in the techniques.

In addition, the best simulations of weightlessness should be developed for ground-based study. The variables and conditions that might cause discrepancies among these studies should be noted and reconciled.

Despite the history of more than 25 years of data gathering and the notable progress that has been made in understanding human responses to weightlessness, questions about the relationship between weightlessness and orthostatic function remain. In addressing these questions, a thorough understanding is required of the various tests of orthostasis, i.e., standing, the "stand test", head-up tilt, and LBNP, including how well the tests compare with each other and with true orthostasis. Furthermore, the mechanisms of post-space-flight and postbedrest intolerances to orthostatic stress should be identified and targeted countermeasures developed.

This information will be valuable clinically and will help prepare humans for the next generation of long-duration space flight.

REFERENCES

1. **Nicogossian, A. E. and Parker, J. F., Jr.,** Space Physiology and Medicine (NASA SP-447), Scientific and Technical Information Branch, National Aeronautics and Space Administration, Washington, D.C. 1982.
2. **Thompson, T. D., Ed.,** *TRW Space Log 1957-1987,* Vol. 23, TRW Space & Technology Group, Redondo Beach, CA, 1988.
3. **Berry, C. A., Coons, D. O., Catterson, A. D., and Kelly, G. F.,** Man's response to long-duration flight in the Gemini spacecraft. in *Gemini Midprogram Conference Including Experiment Results* (NASA SP-121), National Aeronautics and Space Administration, Washington, D.C., 1966, p. 235.
4. **Berry, C. A. and Catterson, A. D.,** Pre-Gemini medical predictions versus Gemini flight results, in: Gemini Summary Conference (NASA SP-138), National Aeronautics and Space Administration, U.S. Government Printing Office, Washington, D.C., 1967, p. 197.
5. **Hawkins, W. R. and Zieglschmid, J. F.,** Clinical aspects of crew health, in *Biomedical Results of Apollo* (NASA SP-368), Johnston, R. S., Dietlein, L. F., and Berry, C. A., Eds., U.S. Government Printing Office, Washington, D.C.,1975, p. 43.
6. **Kerwin, J. P.,** Skylab 2 crew observations and summary, in *Biomedical Results from Skylab* (NASA SP-377), Johnston, R. S. and Dietlein, L. F., Eds., Scientific and Technical Information Office, National Aeronautics and Space Administration, Washington, D.C., 1977, p. 27.
7. **Nicogossian, A. E., LaPinta, C. K., Burchard, E. C., Hoffler, G. W., and Bartelloni, P. J.,** Crew health, in The Apollo-Soyuz Test Project Medical Report (NASA SP-411), Nicogossian, A. E., Ed., Scientific and Technical Information Office, National Aeronautics and Space Administration, Washington, D.C., 1977, p. 11.
8. **Thornton, W. E., Hoffler, G. W., and Rummel, J. A.,** Anthropometric changes and fluid shifts, in Biomedical Results from Skylab (NASA SP-377), Johnston, R. S. and Dietlein, L. F., Eds., Scientific and Technical Information Office, National Aeronautics and Space Administration, Washington, D.C., 1977.
9. **Vorob'yev, Ye. I., Nefedov, Yu. G., Kakurin, L. I., Yegorov, A. D., and Svistunov, I. B.,** Some results of medical investigations made during flights of the "Soyuz-6", "Soyuz-7", and "Soyuz-8" spaceships, *Kosmicheskaya Biologiya i Meditsina,* 4(2), 93, 1970.
10. **Vorob'yev, Ye. I., Yegorov, A. D., Kakurin, L. I., and Nefedov, Yu. G.,** Medical support and principal results of the "Soyuz-9" spaceship crew, *Kosmicheskaya Biologiya i Meditsina,* 4(6), 34, 1970.
11. **Beregovkin, A. V., Krupina, T. N., Syrykh, G. D., Korotayev, M. M., Kuklin, N. A., Balandin, V. A., Znamenskiy, V. S., Nikiforov, V. I., Nistratov, V. V., and Portnov, V. D.,** Results of clinical examination of cosmonauts following a 63-day flight, *Kosmicheskaya Biologiya i Aviakosmicheskaya Meditsina,* 2, 19, 1977.
12. **Molchanov, N. S., Krupina, T. N., Balandin, V. A., Beregovkin, A. V., Korotayev, M. M., Kuklin, N. A., Malyshkin, YeT, Nistratov, V. V., Panfilov, A. S., and Tolstov, V. M.,** Results of clinical examination of A. G. Nikolayev and V. I. Sevast'yanov, *Kosmicheskaya Biologiya i Meditsina,* 4(6), 54, 1970.
13. **Yuganov, Y. M., Degtyarev, V. A., Nekhayer, A. S., et al.,** Dynamics of venous circulation in cosmonauts on the second expedition of Salyut-4, *Kosmicheskaya Biologiya i Aviakosmicheskaya Meditsina,* 2, 31, 1977.
14. **Berry, C. A., Minners, H. A., McCutcheon, E. P., and Pollar, R. A.,** Aeromedical analysis, in Results of the Third Manned Orbital Space Flight, October 3, 1962 (NASA SP-12), National Aeronautics and Space Administration Manned Spacecraft Center, Project Mercury, Scientific and Technical Information Office, National Aeronautics and Space Administration, Washington, D.C., 1962.
15. **Gibson, E. G.,** Skylab 4 crew observations, in Biomedical Results from Skylab (NASA SP-377), Johnston, R. S. and Dietlein, L. F., Eds., Scientific and Technical Information Office, National Aeronautics and Space Administration, Washington, D.C., 1977, 22.
16. **Butusov, A. A., Lyamin, V. R., Lebedev, A. A., Polyakova, A. P., Svistunov, I. B., Tishler, V. A., and Shulenin, A. P.,** Results of routine medical monitoring of cosmonauts during flight of the "Soyuz-9" ship, *Kosmicheskaya Biologiya i Meditsina,* 4(6), 47, 1970.
17. **Vorobyov, E. I., Gazenko, O. G., Gurovsky, N. N., Nefyodov, Yu. G., Egorov, B. B., Bryanov, I. I., Genin, A. M., Degtyarev, V. A., Egorov, A. D., Eryomin, A. V., Kakarin, L. I., Pestov, I. D., and Shulzhenko, F. B.,** Preliminary results of medical investigations during manned flights of the Salyut 4 orbital station, in *Life Sciences and Space Research XV. Proceedings of the Open Meeting of the Working Group on Space Biology,* Pergamon Press, New York, 1977, 199.
18. **Hoffler, G. W., Wolthuis, R. A., and Johnson, R. L.,** Apollo space crew cardiovascular evaluations, *Aerospace Med.,* 45(8), 807, 1974.
19. **Hoffler, G. W. and Johnson, R. L.,** Apollo flight crew cardiovascular evaluations, in Biomedical Results of Apollo (NASA SP-368), Johnston, R. S., Dietlein, L. F., and Berry, C. A., Eds., Scientific and Technical Information Office, National Aeronautics and Space Administration, Washington, D.C., 1975, 227.
20. **Moore, T. P. and Thornton, W. E.,** Inflight and postflight fluid shifts measured by leg volume changes, in Results of the Life Sciences DSOs Conducted Aboard the Space Shuttle 1981-1986, Bungo, M. W., Bagian, T. M., Bowman, M. A., and Levitan, B. M., Space Biomedical Research Institute, Lyndon B. Johnson Space Center, National Aeronautics and Space Administration, Houston, 1987, 59.

21. **Moore, T. P. and Thornton, W. E.,** Space shuttle inflight and postflight fluid shifts measured by leg volume changes, *Aviat., Space, Environ. Med.,* 58(9), A91.

22. **Hoffler, G. W., Bergman, S. A., Jr., and Nicogossian, A. E.,** In-flight lower limb measurement, in The Apollo-Soyuz Test Project Medical Report (NASA SP-411), Nicogossian, A. E., Ed., Scientific and Technical Information Office, National Aeronautics and Space Administration, Washington, D.C., 1977, 63.

23. **Bogomolov, V. V., Popova, I. A., Egorov, A. D., and Kozlovskaya, I. B.,** The results of medical research during the 326-day flight of the second principal expedition on the orbital complex "Mir", in Second US/USSR JWG Conference on Space Biology and Medicine, Gazenko, O. G., Ed., Washington, D.C., 1988.

24. **Kirsch, K. L., Rocker, L., Gauer, O. H., Krause, R., Leach, C., Wicke, H. J., and Landry, R.,** Venous pressure in man during weightlessness, *Science,* 225, 218, 1984.

25. **Kirsch, K. L., Rocker, L., and Haenel, F.,** Venous pressure in space, in Proceedings of the Norderney Symposium on Scientific Results of the German Spacelab Mission D1, Norderney, Germany, 27-29 August 1986, Sahm, P. R., Jansen, R., and Keller, M. H., Eds., 1986, 500.

26. **Gauer, O. H. and Henry, J. P.,** Circulatory basis of fluid volume control, *Physiol. Res.,* 43, 423, 1963.

27. **Gauer, O. H. and Sieker, H. O.,** The continuous recording of central venous pressure changes from an arm vein, *Circ. Res.,* 4, 74, 1956.

28. **Durr, J. A., Simon, C. A., Vallotton, M. B., and Krahenbuhl, B.,** Non-invasive method for measuring central venous pressure, *Lancet,* I, 586, 1978.

29. **Charles, J. B. and Bungo, M. W.,** Noninvasive estimation of central venous pressure using a compact Doppler ultrasound system, in Results of the Life Sciences DSOs Conducted Aboard the Space Shuttle 1981-1986, Bungo, M. W., Bagian, T. M., Bowman, M. A., and Levitan, B. M., Eds., Space Biomedical Research Institute, National Aeronautics and Space Administration, Lyndon B. Johnson Space Center, Houston, 1987, 69.

30. **Baisch, F., Beck, L., Samel, M. D., Samel, A., and Montgomery, L. D.,** Early adaptation of body fluid and cardiac performance to changes in G-level during space flight, Proceedings of the Norderney Symposium on Scientific Results of the German Spacelab Mission D1, Norderney, Germany, 27-29 August 1986, Sahm, P. R., Jansen, R., and Keller, M. H., Eds.,1986, 509.

31. **Alekseyev, D. A., Yarullin, K. K., Krupina, T. N., and Vasil'yeva, T. D.,** Regional hemodynamics in antiorthostatic tests of different intensity, *Kosmicheskaya Biologiya i Aviakosmicheskaya Meditsina,* 8(5), 66, 1974.

32. **Gaffney, F. A., Nixon, J. V., Karlsson, E. S., Campbell, W., Dowdey, A. B. C., and Blomqvist, C. G.,** Cardiovascular deconditioning produced by 20 hours of bedrest with head-down tilt (-5°) in middle-aged healthy men, *Am. J. Cardiol.,* 56, 634, 1985.

33. **Tomaselli, C. M., Kenney, R. A., Frey, M. A. B., and Hoffler, G. W.,** Cardiovascular dynamics during the initial period of head-down tilt. *Aviat. Space Environ. Med.,* 58, 3, 1987.

34. **Nixon, J. V., Murray, R. G., Bryant, C., Johnson, R. L., Jr., Mitchell, J. H., Holland, O. B., Gomez-Sanchez, C., Vergne-Marini, P., and Blomqvist, C. G.,** Early cardiovascular adaptation to simulated zero gravity, *J. Appl. Physiol.,* 46, 541, 1979.

35. **Hargens, A. R.,** Fluid shifts in vascular and extravascular spaces during and after simulated weightlessness, *Med. Sci. Sports Exerc.* 15(5), 421, 1983.

36. **Blomqvist, C. G., Nixon, J. V., Johnson, R. L., Jr., and Mitchell, J. H.,** Early cardiovascular adaptation to zero gravity simulated by head-down tilt, *Acta Astronautica,* 7, 543, 1980.

37. **Lollgen, H., Gebhardt, U., Beier, J., Hordinsky, J., Borger, H., Sarrasch, V., and Klein, K. E.,** Central hemodynamics during zero gravity simulated by head-down bedrest, *Aviat. Space Environ. Med.,* 55, 887, 1984.

38. **Hirasuna, J. D. and Gorin, A. B.,** Effect of prolonged recumbency on pulmonary blood volume in normal humans, *J. Appl. Physiol.,* 50(5), 950, 1981.

39. **Charles, J. B., Bungo, M. W., Ammerman, B., Kreutzberg, K. L., and Youmans, E. M.,** Hemodynamic alterations during the space shuttle prelaunch posture *Aviat. Space Environ. Med.,* 58(5), (Abstr.).491, 1987.

40. **Epstein, M., DeNunzio, A. G., and Ramachandran, M.,** Characterization of renal response to prolonged immersion in man, *J. Appl. Physiol.,* 49(2), 184, 1980.

41. **Risch, W. D., Koubenec, H-J., Gauer, O. H., and Lange, S.,** Time course of cardiac distension with rapid immersion in a thermo-neutral bath, *Pflügers Arch.,* 374, 119, 1978.

42. **Risch, W. D., Koubenec, H-J., Beckmann, U., Lange, S., and Gauer, O. H.,** The effect of graded immersion on heart volume, central venous pressure, pulmonary blood distribution, and heart rate in man, *Pflügers Arch.,* 374, 115, 1978.

43. **Norsk, P., Bonde-Petersen, F., and Warberg, J.,** Arginine vasopressin, circulation, and kidney during graded water immersion in humans, *J. Appl. Physiol.,* 61, 565, 1986.

43a. **Hyatt, K.,** personal communication.

44. **Annat, G., Guell, A., Gauguelin, G., Vincent, M., Mayet, M. H., Bizollon, Ch. A., Legros, J. J., Pottier, J. M., and Gharib, C.,** Plasma vasopressin, neurophysin, renin and aldosterone during a 4-day head-down bed rest with and without exercise, *Eur. J. Appl. Physiol.,* 55, 59, 1986.

45. **Frey, M. A. B., Mader, T., Bagian, J., Charles, J. B., Edwards, B., and Meehan, R.,** Diurnal and duration effects on cardiovascular variables during 48 hours of head-down bedrest, *Aviat. Space Environ. Med.,* 60, 499, 1989.

46. **Katkov, V. Ye, Chestukhin, V. V., Nikolayenko, E. M., Gvozdev, S. V., Rumyantsev, V. V., Guseynova, T. M., and Yegorova, I. A.,** Central circulation in healthy man during 7-day head-down hypokinesia, *Kosmicheskaya Biologiya i Aviakosmicheskaya Meditsina,* 16(5), 45, 1982.

47. **Katkov, V. E., Kakurin, L. I., Chestukhin, V. V., and Kirsch, K.,** Central circulation during exposure to 7-day microgravity, *Physiologist,* 30(1, Suppl.), S36, 1987.

48. **Berry, C. A.,** Aeromedical preparations, in Mercury Project Summary Including Results of the Fourth Manned Orbital Flight May 15 and 16, 1963 (NASA SP-45), National Aeronautics and Space Administration, Manned Spacecraft Center, Project Mercury, Scientific and Technical Information Office, National Aeronautics and Space Administration, Washington, D.C., 1963, 199.

49. **Gazenko, O. G. and Gyurdzhian, A. A.,** The physiological effects of gravitation, in *Problems of Space Biology.* Vol. VI., Sisakyan, N. M., Ed., Nauka Press, Moscow, 1967, 19 (NASA Technical Translation TT F-528, 1969).

50. **Volynkin, Yu. M. and Vasil'yev, P. V.,** Some results of medical studies conducted during the flight of the "Voskhod", in *Problems of Space Biology,* Vol. VI., Sisakyan, N. M., Ed., Nauka, Moscow, 1967, 52 (NASA Technical Translation TT F-528, 1969).

51. **Catterson, A. D., McCutcheon, E. P., Minners, H. A., and Pollard, R. A.,** Aeromedical observations, in Mercury Project Summary Including Results of the Fourth Manned Orbital Flight May 15 and 16, 1963 (NASA SP-45), National Aeronautics and Space Administration, Manned Spacecraft Center, Project Mercury, Scientific and Technical Information Office, National Aeronautics and Space Administration, Washington, D.C., 1963, 299.

52. **Yegorov, A. D., Itsehovskiy, O. G., Alferova, I. V., Turchaninova, V. F., Polenova, A. P., Golubchikova, Z. A., Domracheva, M. V., Lyamin, V. R., and Turbasov, V. D.,** Study of the cardiovascular system (of Salyut-6 prime crews), in *Results of Medical Research Performed on Board the Salyut-6 - "Soyuz" Orbital Scientific Research Complex,* Gurovskiy, N. N., Ed., Nauka, Moscow, 1986, 89.

53. **Turbasov, V. D., Golubchikova, Z. A., Lyamin, V. R., and Romanov, Ye. M.,** Results of electrocardiographic examinations of Salyut-7 Soyuz prime crews, in *Kosmicheskaya Biologiya i Aviakosmicheskaya Meditsina,* Gazenko, O. G., Ed.(abstracts of papers delivered at the Eighth All-Union Conference), Nauka, Moskow, 1986, 143.

54. **Bungo, M. W. and Johnson, P. C., Jr.,** Cardiovascular examinations and observations of deconditioning during the space shuttle orbital flight test program, *Aviat. Space Environ. Med.,* 54(11), 1001, 1983.

55. **Bungo, M. W., Goldwater, D. J., Popp, R. L., and Sandler, H.,** Echocardiographic evaluation of space shuttle crewmembers, *J. Appl. Physiol.,* 62(1), 278, 1987.

56. **Johnson, R. L., Hoffler, G. W., Nicogossian, A. E., Bergman, S. A., and Jackson, M. M.,** Lower body negative pressure: third manned Skylab mission, in Biomedical Results from Skylab (NASA SP-377), Johnston, R. S. and Dietlein, L. F., Eds, Scientific and Technical Information Office, National Aeronautics and Space Administration, Washington, D.C., 1977, 284.

56a. **Thornton, W.,** personal communication.

57. **Bungo, M. W., Charles, J. B., and Johnson, P. C., Jr.,** Cardiovascular deconditioning during space flight and the use of saline as a countermeasure to orthostatic intolerance. *Aviat. Space Environ. Med.,* 56, 985, 1985.

58. **Henry, W. L., Epstein, S. E., Griffith, J. M., Goldstein, R. E., and Redwood, D. R.,** Effect of prolonged space flight on cardiac function and dimensions, in Biomedical Results from Skylab (NASA SP-377), Johnston, R. S. and Dietlein, L. F., Eds., Scientific and Technical Information Office, National Aeronautics and Space Administration, Washington, D.C., 1977, 366.

59. **Charles, J. B. and Bungo, M. W.,** Post-space flight changes in resting cardiovascular parameters are associated with preflight left ventricular volume, *Aviat. Space Environ. Med.,* 57 (Abstr.):493, 1986.

60. **At'kov, O. Yu.,** The state of cosmonauts' cardiovascular systems during long-term orbital flights, *Byulleten' Vsesoyuznogo Kardiologicheskogo Nauchnogo Tsentra AMN SSSR,* 8(2), 97, 1985.

61. **At'kov, O. Yu., Bednenko, V. S., and Fomina, G. A.,** Ultrasound techniques in space medicine, *Aviat. Space Environ. Med.,* 58 (9, Suppl.), A69, 1987.

62. **Pourcelot, L., Savilov, A. A., Bystrov, V. V., Kakurin, L. I., Kotovskaya, A. R., Patat, F., Pottier, J. M., and Zhernakov, A. F.,** Results of echocardiographic examination during 7 days onboard Salyut VII, June 1982. *Physiologist,* 26 (6, Suppl.), S66, 1983.

63. **Pourcelot, L., Arbeille, P., Pottier, J.-M., Patat, F., Berson, M., Roncin, A., LeToullec, C., Charles, J., Guell, A., and Gharib, C.,** Results of cardiovascular examination during 51-G mission, in Physiologic Adaptation of Man in Space, 7th International Man in Space Symposium, Space Adaptation, abstracts of papers, Houston, Texas, February 10-13, 1986.

64. **Bungo, M. W, Charles, J. B., Riddle, J., Roesch, J. A., Wolf, D. A., and Seddon, R.,** Echocardiographic investigation of the hemodynamics of weightlessness, 35th Annual Scientific Session, American College of Cardiology. *J Am. Coll. Cardiol.,* 7(2) (Abstr.) :192A, 1986.

65. **Hoffler, G. W., Nicogossian, A. E., Bergman, S. A., Jr., and Johnson, R. L.,** Cardiovascular evaluations, in The Apollo-Soyuz Test Project Medical Report (NASA SP-411), Nicogossian A. E., Ed., Scientific and Technical Information Office, National Aeronautics and Space Administration, Washington, D.C., 1977, 59.

66. **Nicogossian, A. E., Hoffler, G. W., Johnson, R. L., and Gowen, R. J.,** Determination of cardiac size from chest roentgenograms following Skylab missions, in Biomedical Results from Skylab (NASA SP-377), Johnston, R. S. and Dietlein, L. F., Eds., Scientific and Technical Information Office, National Aeronautics and Space Administration, Washington, D.C., 1977, 400.

67. **Volicer, L., Jean-Charles, R., and Chobanian, A. V.,** Effects of head-down tilt on fluid and electrolyte balance, *Aviat. Space Environ. Med.,* 47, 1065, 1976.

68. **Knitelius, H. and Stegemann, J.,** Heart volume during short-term head-down tilt (-6°) in comparison with horizontal body position, *Aviat. Space Environ. Med.,* 58 (9, Suppl.), A61, 1987.

69. **Bonde-Petersen, F., Suzuki, Y., and Sadamoto, T.,** Cardiovascular effects of simulated zero-gravity in humans, *Acta Astronautica,* 10(9), 657, 1983.

70. **Saltin, B., Blomqvist, G., Mitchell, J. H., Johnson, R. L., Jr., Wildenthal, K., and Chapman, C. B.,** Response to exercise after bedrest and after training, *Circulation,* 38 (5, Suppl. 7), 1, 1968.

71. **Katkov, V. E., Chestukhin, V. V., Nikolayenko, E. M., Rumyantsev, V. V., and Gvozdev, S. V.,** Central circulation of a normal man during 7-day head-down tilt and decompression of various body parts, *Aviat. Space Environ. Med.,* 54 (12, Suppl. 1), S24, 1983.

72. **Maksimov, D. G. and Domracheva, M. V.,** Changes in central and peripheral hemodynamics during prolonged antiorthostatic hypokinesia as weightlessness models, *Kosmicheskaya Biologiya i Aviakosmicheskaya Meditsina,* 10(5), 52, 1976.

73. **Chobanian, A. V., Lille, R. D., Tercyak, A., and Blevins, P.,** The metabolic and hemodynamic effects of prolonged bed rest in normal subjects, *Circulation,* 49, 551, 1974.

74. **Guell, A., Bes, A., Braak, L., and Barrere, M.,** Effects of a weightlessness simulation on the velocity curves measured by Doppler sonography at the level of the carotid system, *Physiologist,* 22 (Suppl.), S25, 1979.

75. **Guell, A., Dupui, P.H., Barrere, M., Fanjaud, G., and Bes, A.,** Changes in the loco-regional cerebral blood flow (r.C.B.F.) during a simulation of weightlessness, *Acta Astronautica,* 9(11), 689, 1982.

76. **Artamonova, N. P., Zakharova, T. S., Morukov, B. V., Arzamazov, G. S., and Semenov, V. Yu.,** Bioelectrical activity of the heart and blood electrolytes in essentially healthy subjects submitted to anti-orthostatic hypokinesia for 120 days, *Kosmicheskaya Biologiya i Aviakosmicheskaya Meditsina,* 20(6), 34, 1986.

77. **Kuz'min, M. P.,** Reactions of eye retinal vessels and intraocular pressure during man's 120-day restriction to a horizontal posture, *Kosmicheskaya Biologiya i Aviakosmicheskaya Meditsina,* 7(2), 98, 1973.

78. **Stremel, R. W., Convertino, V. A., Bernauer, E. M., and Greenleaf, J. E.,** Cardiorespiratory deconditioning with static and dynamic leg exercise during bed rest, *J. Appl. Physiol.,* 41(6), 905, 1976.

79. **Greenleaf, J. E., Wade, C. E., and Leftheriotis, G.,** Orthostatic responses following 30-day bedrest deconditioning with isotonic and isokinetic exercise training, *Aviat. Space Environ. Med.,* 60, 537, 1989.

80. **Kakurin, L. I., Katkovskiy, B. S., Tishler, V. A., Kozyrevskaya, G. I., Shashkov, V. S., Georgiyevskiy, V. S., Grigor'yev, A. I., Mikhaylov, V. M., Anashkin, O. D., Machinskiy, G. V., Savilov, A. A., and Tikhomirov, Ye. P.,** Substantiation of a set of preventive measures referable to the objectives of missions in the Salyut orbital station, *Kosmicheskaya Biologiya i Aviakosmicheskaya Meditsina,* 3, 20, 1978.

81. **Convertino, V. A., Sandler, H., Webb, P., and Annis, J. F.,** Induced venous pooling and cardiorespiratory responses to exercise after bed rest, *J. Appl. Physiol.* 52(5), 1343, 1982.

82. **Stevens, P. M., Miller, P. B., Lynch, T. N., Gilbert, G. A., Johnson, R. L., and Lamb, L. E.,** Effects of lower body negative pressure on physiologic changes due to four weeks of hypoxic bed rest, *Aerospace Med.,* 37, 466, 1966.

83. **Kakurin, L. I.,** Peculiarities in the function of human cardiorespiratory system during a six-month anti-orthostatic hypokinesia, in XII US/USSR Joint Working Group Meeting on Space Biology and Medicine, Washington, D.C., November 9-22, 1981.

84. **Goldberger, A. L., Goldwater, D., and Bhargava, V.,** Atropine unmasks bed-rest effect: a spectral analysis of cardiac interbeat intervals, *J. Appl. Physiol.,* 61(5), 1843, 1986.

85. **Shulzhenko, E. B., Tigranyan, R. A., Panfilov, V. E., and Bzhalava, I. I.,** Physiological reactions during acute adaptation to reduced gravity, in *Life Sciences and Space Research, Vol. 18. Proceedings of the Open Meeting of the Working Group on Space Biology,* Pergamon Press, Elmsford, NY, 1980, 175.

86. **Leach, C. S.,** Review of endocrine results: Project Mercury, Gemini Program, and Apollo Program, in, *Proceedings of the 1970 Manned Spacecraft Center Endocrine Program Conference,* (NASA Technical Memorandum TM X-58068), Preventive Medicine Division, Medical Research and Operations Directorate, National Aeronautics and Space Administration, Washington, D.C.,1971, 3-1.

87. **Thornton, W. E. and Ord, J.,** Physiological mass measurements in Skylab, in Biomedical Results from Skylab (NASA SP-377), Johnston, R. S. and Dietlein, L. F., Eds, Scientific and Technical Information Office, National Aeronautics and Space Administration, Washington, D.C., 1977, 175.

88. **Dietlein, L. F. and Harris, E.,** Experiment M-5, bioassays of body fluids, in *Gemini Midprogram Conference Including Experiment Results* (NASA SP-121), National Aeronautics and Space Administration, Washington, D.C., 1966, 403.

89. **Leach, C. S., Alexander, W. C., and Johnson, P. C.,** Endocrine, electrolyte, and fluid volume changes associated with Apollo missions, in Biomedical Results of Apollo (NASA SP-368), Johnston, R. S., Dietlein, L. F., and Berry, C. A., Eds., U.S. Government Printing Office, Washington, D.C., 1975, 163.

90. **Grigoriev, A. I., Popova, I. A., and Ushakov, A. S.,** Metabolic and hormonal status of crewmembers in short-term spaceflights, *Aviat. Space Environ. Med.,* 58 (9, Suppl.), A121, 1987.

91. **Leach, C. S. and Rambaut, P. C.,** Biochemical Responses of the Skylab crewmen, in Biomedical Results from Skylab (NASA SP-377), Johnston, R. S. and Dietlein, L. F., Eds., Scientific and Technical Information Office, National Aeronautics and Space Administration, Washington, D.C., 1977, 204.

92. **Leach, C. S., Inners, L. D., and Charles, J. B.,** Changes in total body water during space flight, in Results of the Life Sciences DSOs Conducted Aboard the Space Shuttle 1981-1986, Bungo, M. W., Bagian, T. M., Bowman, M. A., and Levitan, B. M, Space Biomedical Research Institute, Lyndon B. Johnson Space Center, National Aeronautics and Space Administration, Houston, 1987, 49.

93. **Leach, C. S.,** Biochemistry and endocrinology results, in The Apollo-Soyuz Test Project Medical Report (NASA SP-411), Nicogossian, A.E., Ed., Scientific and Technical Information Office, National Aeronautics and Space Administration, Washington, D.C., 1977, 87.

94. **Johnson, P. C., Driscoll, T. B., and LeBlanc, A. D.,** Blood volume changes, in Biomedical Results from Skylab (NASA SP-377), Johnston, R. S. and Dietlein, L. F., Eds., Scientific and Technical Information Office, National Aeronautics and Space Administration, Washington, D.C., 1977, 235.

95. **Noskov, V. B., Afonin, B. V., Levedev, V. I., Boyka, P. A., Sukhanov, Yu. V., Kravchenko, V. V., and Kvetyanski, R.,** Fluid-electrolyte metabolism and its hormonal regulation under conditions of long-term space flight, in *Kosmicheskaya Biologiya i Aviakosmicheskaya Meditsina,* Gazenko, O. G., Ed., (abstracts of papers delivered at the Eighth All-Union Conference) Nauka, Moskow, 1986, 9.

96. **Leach, C. S.,** Medical results from STS 1-4: analysis of body fluids, *Aviat. Space Environ. Med.,* 54 (12, Suppl. 1), S50, 1983.

97. **Leach, C. S.,** Fluid control mechanisms in weightlessness, *Aviat. Space Environ. Med.,* 58 (9, Suppl.), A74, 1987.

98. **Davydova, N. A., Tigranian, R. A., Kvetnansky, R., and Macho, L.,** Reactions of the sympatho-adrenal system of cosmonauts after space flights of varying duration, in *Stress: The Role of Catecholamines and Other Neurotransmitters II,* Usdin, E., Kvetnansky, R., and Axelrod, J., Eds., Gordon and Breach, New York, 1983, 977.

99. **Tigranian, R. A., Kalita, N. F., Macho, L., Kvetnansky, R., and Vigas, M.,** Changes of some pituitary hormones in cosmonauts after space flights of different duration, in *Stress: The Role of Catecholamines and Other Neurotransmitters II,* Usdin, E., Kvetnansky, R., and Axelrod, J., Eds., Gordon and Breach , New York, 1983, 1003.

100. **Dallman, M. F., Vernikos, J., Keil, L. C., O'Hara, D., and Convertino, V.,** Hormonal, fluid, and electrolyte responses to 6° anti-orthostatic bedrest in healthy male subjects, in *Stress: The Role of Catecholamines and Other Neurotransmitters II,* Usdin, E., Kvetnansky, R., and Axelrod, J., Eds., Gordon and Breach, New York, 1983, 1057.

101. **Norsk, P., Bonde-Petersen, F., and Warberg, J.,** Central venous pressure and plasma arginine vasopressin in man during water immersion combined with changes in blood volume, *Eur. J. Appl. Physiol.,* 54, 608, 1986.

102. **Miki, K., Shiraki, K., Sagawa, S., DeBold, A. J., and Hong, S. K.,** Atrial natriuretic factor during head-out immersion at night, *Am. J. Physiol.,* 254, R235, 1988.

103. **Graveline, D. E. and Barnard, G. W.,** Physiologic effects of a hydrodynamic environment: short-term studies, *Aerospace Med.,* 32, 726, 1961.

104. **Shiraki, K., Konda, N., Sagawa, S., Claybaugh, J. R., and Hong, S. K.,** Cardiorenal-endocrine responses to head-out immersion at night, *J. Appl. Physiol.,* 60(1), 176, 1986.

105. **Greenleaf, J. E., Shvartz, E., Kravik, S., and Keil, L. C.,** Fluid shifts and endocrine responses during chair rest and water immersion in man, *J. Appl. Physiol.,* 48(1), 79, 1980.

106. **Khosla, S. S. and DuBois, A. B.,** Osmoregulation and interstitial fluid pressure changes in humans during water immersion, *J. Appl. Physiol.,* 51(3), 686, 1981.

107. **Epstein, M.,** Water immersion and the kidney: implications for volume regulation, *Undersea Biomed. Res.,* 11(2), 113, 1984.

108. **Gerbes A. L., Arendt, R. M., Gerzer, R., Schnizer, W., Jungst, D., Paumgartner, G., and Wernze, H.,** Role of atrial natriuretic factor, cyclic GMP and the renin-aldosterone system in acute volume regulation of healthy human subjects, *Eur. J. Clin. Invest.,* 18, 425, 1988.

108a. **Greenleaf, J.,** personal communication.

109. **Mano, T., Yamazaki, Y., Mitarai, G., and Iwase, S.,** Somatosensory-vestibular interactions and autonomic nervous activities in man under weightlessness simulated by water immersion, in *Sensory-Motor Functions Under Weightlessness and Space Motion Sickness,* Proceedings of the International Symposium on Space Medicine, Mitarai, G. and Igaraski, M., Eds., University of Nagoya Press, Nagoya, 1985, 1.

110. **Epstein, M., Johnson, G., and DeNunzio, A. G.,** Effects of water immersion on plasma catecholamines in normal humans, *J. Appl. Physiol.,* 54(1), 244, 1983.

111. **Greenleaf, J. E., van Beaumont, W., Bernauer, E. M., Haines, R. D., Sandler, H., Staley, R. W., Young, H. L., and Yusken, J. W.,** Effects of rehydration on +Gz tolerance after 14-days bed rest, *Aerospace Med.,* 44, 715, 1973.

112. **Stevens, P. M., Lynch, T. N., Johnson, R. L., and Lamb, L. E.,** Effects of 9-alphaflurohydrocortisone and venous occlusive cuffs on orthostatic deconditioning of prolonged bed rest, *Aerospace Med.,* 37, 1049, 1966.

113. **Hyatt, K. H., Jacobson, L. B., and Schneider, V. S.,** Comparison of 70° tilt, LBNP, and passive standing as measures of orthostatic tolerance, *Aviat. Space Environ. Med.,* 46(6), 801, 1975.

114. **Hyatt, K. H., Kamenetsky, L. G., and Smith, W. M.,** Extravascular dehydration as an etiologic factor in post-recumbency orthostatism, *Aerospace Med.,* 40(6), 644, 1969.

115. **Dill, D.B. and Costill, D.L.,** Calculation of percentage changes in volumes of blood, plasma, and red cells in dehydration, *J. Appl. Physiol.,* 37, 247, 1974.

116. **Johnson, P. C.,** Fluid volumes changes induced by space flight, *Acta Astronautica,* 6, 1335, 1979.

117. **Hyatt, K. H. and West, D. A.,** Reversal of bedrest-induced orthostatic intolerance by lower body negative pressure and saline, *Aviat. Space Environ. Med.,* 48(2), 120, 1977.

118. **Beckett, W. S., Vroman, N. B., Nigro, D., Thompson-Gorman, S., Wilkerson, J. E., and Fortney, S. M.,** Effect of prolonged bed rest on lung volume in normal individuals, *J. Appl. Physiol.,* 61(3), 919, 1986.

119. **Fortney, S. M., Beckett, W. S., Carpenter, A. J., Davis, J., Drew, H., LaFrance, N. D., Rock, J. A., Tankersley, C. G., and Vroman, N. B.,** Changes in plasma volume during bedrest: effects of menstrual cycle and estrogen administration, *J. Appl. Physiol.,* 65(2), 525, 1988.

120. **Sandler, H., Webb, P., Annis, J., Pace, N., Grunbaum, B. W., Dolkas, D., and Newsom, B.,** Evaluation of a reverse gradient garment for prevention of bed-rest deconditioning, *Aviat. Space Environ. Med.,* 54(3), 191, 1983.

121. **Greenleaf, J. E.,** Mechanisms for negative water balance during weightlessness; an hypothesis, *J. Appl. Physiol.,* 60(1), 60, 1986.

122. **Eckberg, D. L., Charles, J. B., Fritsch, J. M., Bennett, B. S., and Bungo, M. W.,** Effects of short-duration space flight on human carotid baroreceptor cardiac reflex responses (abstract), in 8th International Academy of Astronautics Man in Space Symposium, Tashkent, Uzbekistan, USSR, September 29-October 3, 1989.

123. **Convertino, V. A., Doerr, D. F., Eckberg, D. L., Fritsch, J. M., and Vernikos-Danellis, J.,** Carotid baroflex response following 30 days exposure to simulated microgravity, *Physiologist,* 32 (Suppl.), 5, 1989.

124. **Charles, J. B. and Bungo, M. W.,** Cardiac dimensions and orthostatic heart rate as a function of time in microgravity (abstract), in 7th Man in Space Symposium, Houston, TX. February 10-13, 1986.

125. **Kotovskaya, A. R., Vartbaranov, R. V., and Simpura, S. F.,** Human physiological reactions during the action of transverse accelerations following hypodynamia, in *Problems of Space Biology,* Vol. VI, Sisakyan, N. M., Ed., Nauka, Moscow, 1967, 107 (NASA Technical Translation TT F-528, 1969).

126. **Volynkin, Yu. M. and Vasil'yev, P. V.,** Some results of medical studies conducted during the flight of the "Voskhod", in *Problems of Space Biology,* Vol. VI, Sisakyan, N. M., Ed., Nauka Press, Moscow, 1967, 52 (NASA Technical Translation TT F-528, 1969).

127. **Petukhov, B. N., Purakhin, Yu. N., Georgiyevskiy, V. S., Mikhaylov, V. M., Smyshlyayeva, V. V., and Fat'yanova, L. I.,** Regulation of erect posture of cosmonauts after an 18-day orbital flight, *Kosmicheskaya Biologiya i Meditsina,* 4(6), 50, 1970.

128. **Kalinichenko, V. V., Gornago, V. A., Machinskiy, G. V., Zhelgurova, Yu. D., Pometov, Yu, D., and Katkovskiy, B. S.,** Dynamics of orthostatic stability of cosmonauts after flight aboard the "Soyuz-9" spaceship, *Kosmicheskaya Biologiya i Meditsina,* 4(6), 68, 1970.

129. **Gazenko, O. G., Shulzhenko, E. B., and Egorov, A. D.,** Cardiovascular changes in prolonged space flights (preprint), in 36th Congress of the International Astronautical Federation, (IAF-85-323), Oxford, Pergamon Press, 1985.

130. **Degtyarev, V. A., Doroshev, V. G., Kirillova, Z. A., Lapshina, N. A., Ponomarev, S. I., and Ragozin, V. N.,** Reactions to LBNP test of the crew of the Salyut-5 orbital station, *Kosmicheskaya Biologiya i Aviakosmicheskaya Meditsina,* 14(4), 11, 1980.

131. **Tomaselli, C. M., Frey, M. A. B., Kenney, R. A., and Hoffler, G. W.,** Effect of a central redistribution of fluid volume on response to lower body negative pressure, *Aviat. Space Environ. Med.,* in press.

132. **Lollgen, H., Klein, K. E., Gebhardt, U., Beier, J., Hordinsky, J., Sarrasch, V., Borger, H., and Just, H.,** Hemodynamic response to LBNP following 2 hours head-down tilt (-6°), *Aviat. Space Environ. Med.,* 57, 406, 1986.

133. **Hordinsky, J. R., Gebhardt, U., Wegmann, H. M., and Schafer, G.,** Cardiovascular and biochemical response to simulated space flight entry, *Aviat. Space Environ. Med.,* 52(1), 16, 1981.

134. **Dietrick, J. E., Whedon, G. D., and Shorr, E.,** Effects of immobilization upon various metabolic and physiologic functions of normal men, *Am. J. Med.,* 4, 3, 1948.

135. **Goldwater, D., Montgomery, L., Hoffler, G. W., Sandler, H., and Popp, R.,** Echocardiographic and peripheral vascular responses of men (ages 46 to 55) to lower body negative pressure (LBNP) following 10 days of bed rest, Preprints of Annual Scientific Meeting, Aerospace Medical Association, 1979, 51.

136. **Muller, E. W., Hohlweck, H., Plath, G., and Baisch, F.,** Leg volume changes. Responses to LBNP during 7 days of 0-G simulation (6° HDT), in Second European Symposium of Life Sciences Research in Space, Porz Wahn, Germany, June 4-6, 1984 (ESA SP-212, August 1984), 162.

137. **Greenleaf, J. E., Haines, R. F., Bernauer, E. M., Morse, J. T., Sandler, H., Armbruster, R., Sagan, L., and van Beaumont, W.,** +Gz tolerance in man after 14-day bedrest periods with isometric and isotonic exercise conditioning, *Aviat. Space Environ. Med.,* 46(5), 671, 1975.

138. **Beregovkin, A. V. and Kalinichenko, V. V.,** Reactions of the cardiovascular system during 30-day simulation of weightlessness by means of antiorthostatic hypokinesia, *Kosmicheskaya Biologiya i Aviakosmicheskaya Meditsina,* 8(1), 72, 1974.

138a. **Smith, J. J.,** personal communication.

139. **Kakurin, L. I., Lobachik, V. I., Mikhailov, V. M., and Senkevich, Yu. A.,** Antiorthostatic hypokinesia as a method of weightlessness simulation, *Aviat. Space Environ. Med.,* 47(10), 1083, 1976.

140. **Sandler, H.,** Cardiovascular responses to weightlessness and ground-based simulations, in *Zero-g Simulation for Ground-Based Studies in Human Physiology, with Emphasis on the Cardiovascular and Body Fluid Systems,* European Space Agency, 1982, 107.

141. **Vil-Vil'yams, I. F. and Shulzhenko, Ye. B.,** Functional state of the cardiovascular system under the combined effect of 28-day immersion, rotation on a short-arm centrifuge and exercise on a bicycle ergometer, *Kosmicheskaya Biologiya i Aviakosmicheskaya Meditsina,* 14(2), 42, 1980.

142. **Shulzhenko, Ye. B., Vil-Vil'yams, I. F., and Panfilov, V. E.,** Human adaptation to simulated gravitational fields, *Acta Astronautica,* 9(3), 173, 1982.

143. **Blomqvist, C. G., Gaffney, F. A., and Nixon, J. V.,** Cardiovascular responses to head-down tilt in young and middle-aged men, *Physiologist,* 26 (Suppl.), S81, 1983.

144. **Miller, P. B. and Leverett, S. D.,** Tolerance to transverse (+Gx) and headward (+Gz) acceleration after prolonged bed rest, *Aerospace Med.,* 36, 13, 1965.

145. **van Beaumont, W., Greenleaf, J. E., and Juhos, L.,** Disproportional changes in hematocrit, plasma volume, and proteins during exercise and bed rest, *J. Appl. Physiol.,* 33, 55, 1972.

146. **Sandler, H. and Winter, D. L.,** Physiological Responses of Women to Simulated Weightlessness (NASA SP-430), National Aeronautics and Space Administration, Washington, D.C., 1978.

147. **Beck, L. and Baisch, F.,** Non-invasive assessment of heart contractility changes during a 7 day -6° head-down tilt 0-G simulation, *Physiologist,* 27 (6 Suppl.), S57, 1984.

148. **Convertino, V. A. and Sandler, H.,** VO$_2$ kinetics during submaximal exercise following simulated weightlessness, *Physiologist,* 25 (6, Suppl), S159, 1982.

149. **Stegemann, J., Essfeld, D., and Hoffman, U.,** Effects of a 7-day head-down tilt (-6°) on the dynamics of oxygen uptake and heart rate adjustment in upright exercise, *Aviat. Space Environ. Med.,* 56, 410, 1985.

150. **Convertino, V. A., Bisson, R., Bates, R., Goldwater, D., and Sandler, H.,** Effects of antiorthostatic bedrest on the cardiorespiratory responses to exercise, *Aviat. Space Environ. Med.,* 52, 251, 1981.

151. **Montgomery, L. D.,** Body volume changes during simulated weightlessness: an overview, *Aviat. Space Environ. Med.,* 58 (Suppl.), A80, 1987.

152. **Polese, A., Goldwater, D., London, L., Yuster, D., and Sandler, H.,** Resting cardiovascular effects of horizontal (0°) and head-down (-6°) bedrest on normal men, in Preprints of the Scientific Program, Annual Meeting of the Aerospace Medical Association, 1980, 24.

153. **Sandler, H.,** Effects of bedrest and weightlessness on the heart, in *Hearts and Heart-like Organs,* Vol. 2, Academic Press, New York, 1980.

154. **Greenleaf, J. E., Greenleaf, C. J., van Derveer, D., and Dorchak, K. J.,** Adaptation to prolonged bedrest in man: a compendium of research, (NASA TM X-3307), National Aeronautics and Space Association Washington, D.C., 1976.

155. **Greenleaf, J. E., Silverstein, L., Bliss, J., Largerheison, V., Rossow, H., and Chao, C.,** Physiological responses to prolonged bedrest and fluid immersion in man: a compendium of research (1974-1980) (NASA TM 81324), National Aeronautics and Space Administration Washington, D.C., 1982.

156. **Hoffler, G. W.,** Cardiovascular studies of U.S. space crews: an overview and perspective, in *Cardiovascular Flow Dynamics and Measurements,* Hwang, N. H. C. and Norman, N. A., Eds., University Park Press, Baltimore, MD, 1977.

156a. **Coons, D. O.,** personal communication.

Chapter 4

AGE AND THE RESPONSE TO ORTHOSTATIC STRESS

James J. Smith and Carol J. M. Porth

TABLE OF CONTENTS

I. GENERAL CIRCULATORY CHANGES WITH AGE

With the steady increase in the over-65 population, now about 12% in the United States, and the steep rise in the incidence of cardiovascular disease in this age group, increasing attention has been focused on cardiovascular function in the elderly and factors that influence it. Not only does the incidence of circulatory disease increase markedly with age, but there is also a striking similarity between the effects of age and early coronary heart disease. This sometimes poses a serious problem in differential diagnosis and makes it particularly important to establish norms for cardiovascular function in the aged before reaching conclusions regarding cardiovascular pathology. Since the ability of the body to respond effectively to postural change is importantly influenced by the general circulatory changes with aging, these changes will be briefly reviewed in this section.

A. STRUCTURAL AND FUNCTIONAL CHANGES IN THE CIRCULATION

Some of these circulatory changes that occur with aging are important and have far-reaching physiological consequences. The decreased arterial distensibility and decreased venous compliance with age occur in both men and animals, and are related to a progressive increase in the collagen/elastin ratio of the vessel wall. There is increasing stiffness of the aorta and large arteries in both a radial and longitudinal direction, as determined by pressure-volume curves.[1,3]

This increasing arterial rigidity is a diffuse process involving the replacement of elastin with connective tissue; the result is due to an alteration of the viscoelastic properties of the wall at large and not to atherosclerosis, which is generally a patchy and irregular process. While the increase in aortic volume, especially after age 60, provides some hemodynamic buffering of the arterial pressure-volume curve, this is quite inadequate to offset the marked increase in stiffness in an older aorta. This age effect is illustrated by the fact that in a 70-year-old aorta, a 100-mmHg rise in aortic pressure is achieved with an aortic volume increase of only half that required in a 20-year-old aorta.[3]

This decreased aortic distensibility, along with the increased peripheral vascular resistance of the arteriolar bed, serves to increase the mechanical impedance of the aorta and impose a heavier afterload on the aging left ventricle. This increased impedance to ejection is very likely an important factor in limiting the increase in left ventricular stroke volume during exercise in the older subject. Associated with the increased aortic impedance, there is also a left ventricular hypertrophy, a modest increase in left ventricular chamber size, and, very importantly, a decreased left ventricular compliance. In many healthy older subjects, these physical changes result in a slowing of left ventricular relaxation and contraction. This results in a limitation of end-diastolic volume, a prolongation of left ventricular ejection time, and a limited Frank-Starling response. There is also in older subjects, increased stiffness of the heart valves and pericardium, which may exert a functional handicap.[3]

Aside from these alterations in the cardiovascular system, each step in the path that oxygen takes from the outer air to the metabolizing cell is vulnerable to aging. The pulmonary alveoli become smaller and more shallow, so there is a steady decline in vital capacity and about a 20% decrease in pulmonary alveolar surface area by age 70. There is a decrease in pulmonary elastic recoil and lung compliance, and because of the stiffened rib cage, there is increasing reliance on diaphragmatic breathing. Since skeletal muscle typically shows prolonged contraction and relaxation periods, all the timed, ventilatory functions are slower.[4]

The alveolar-arterial O_2 difference is increased with age. Although hemoglobin (Hb) concentration is maintained in the healthy older person, there is a small alteration in red blood cell metabolism and a resultant decreased concentration of 2,3-diphosphoglycerate. As a consequence, there is a shift of the HbO_2 dissociation curve to the left, thereby making the tissue unloading of O_2 more difficult.[4]

B. AGE EFFECTS ON THE RESTING CIRCULATION

There is in older individuals a gradual decline in the resting values of cardiac index (CI), stroke volume index (SVI), and arteriovenous (A-V) O_2 difference, and generally a steady rise in systolic and diastolic pressure and peripheral vascular resistance.[1,2] Perhaps the most significant circulatory fact of aging is the steady decrease of cardiac output, both at rest and during exercise. Since the most important determinant of cardiac output is oxygen metabolism, it is the steady decline of the basal metabolic rate that is the fundamental cause of the fall in cardiac output. The output of the heart decreases at a rate of about 230 ml/m^2/min per decade. This means a decrease of about one third from age 20 to age 80.[5] The metabolism decrease is probably due to a lessening in the efficiency of the tissue oxidative enzymes as manifested by a steady decrease in the resting A-V O_2 difference.[6,7] It is most likely that the heart, if healthy, plays mainly a permissive role in the circulatory system and is not the primary cause of the declining cardiac output. The distribution of the diminished cardiac output is not an equitable one; cerebral and coronary blood flow change little with age, but renal and splanchnic flow decrease significantly.[2,3]

It should be pointed out that the cardiovascular changes described above were average findings from large samplings of different age populations. Physiological age often does not parallel chronological age, and there frequently are wide individual differences in functional capability. For example, in some individuals, arterial pressures do not rise with age, and many of these subjects do not have increased systemic vascular resistance.[6] Such differences will depend mainly upon physical condition and hereditary, dietary, and socioenvironmental factors.

C. AGE EFFECTS ON EXERCISE RESPONSE

While the changes described above impose relatively little handicap to the normal, older individual at rest, circulatory stresses such as exercise will uncover the shallowness of the physiological reserve and will reveal significant functional deficiencies with increasing age. Probably the most important of these is the decline with age of the maximum O_2 consumption (Vo_{2max}). In adults, Vo_{2max} steadily decreases by about 1% per year, so that at age 65 it is about 60% of its value at age 20.[8] Thus, dynamic work capacity (i.e., Vo_{2max} per unit body weight) is severely restricted with age. Older men and women respond to exercise with progressively lesser increases in heart rate (HR), stroke volume, cardiac output, and A-V O_2 difference and with greater increases in systolic pressure, all of which distinctly limit the exercise response. There is also evidence that in many elderly persons, there is, with exercise, a decreased ventricular ejection fraction, suggesting diminished ventricular contractility.[9]

A recent study has shown that sedentary young men have about 30% greater Vo_{2max} than sedentary young women, but the rate of decline with age was significantly greater in the men than in the women. The sedentary young men also had higher HR_{max}, cardiac index$_{max}$, and SVI_{max} values than the young women, but again (except in the case of the SVI), the rate of decrease was greater in the sedentary men. The decline of Vo_{2max} is also more rapid in the obese and in the smoker.[3]

It has been consistently observed that in humans, peripheral vascular resistance at rest and at comparable workloads is increased with age and that in exercise there is a lesser dilator capability of vascular smooth muscle. It is also significant that during exercise, the elderly have elevated pulmonary arterial wedge pressures and increased right ventricular end-diastolic pressures, suggesting a diminished reserve capacity in older hearts. Generally, the attainment of steady-state levels of HR, blood pressure, and ventilatory responses is slower and the rate of recovery more prolonged with increasing age.[2,3,6]

The mechanism of these changes in exercise response is not certain. Some investigators believe the causes are primarily metabolic, i.e., a diminished capacity of the tissue oxidative enzymes.[6,7] Others believe that the structural and functional defects of the circulatory system itself are primarily responsible. It is likely that in many instances both mechanisms are operative.

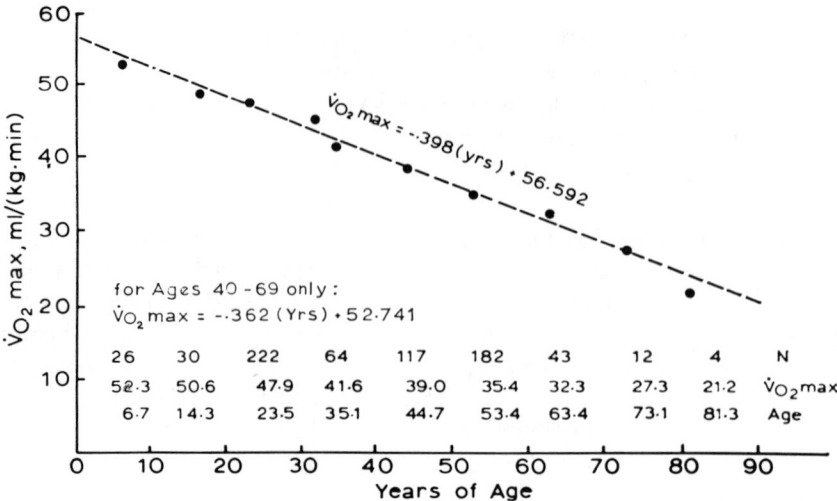

FIGURE 1. Decrease of Vo_{2max} with age as determined in 700 men and boys, corrected for body weight. Decline is steady at about 1% per year so that at age 65, Vo_{2max} is about 60% of its value at age 20. (From Dehn, M. M. and Bruce, R. A., *J. Appl. Physiol.,* 33, 805, 1972. With permission.)

Certainly, the evidence would suggest that the diminishing cardiovascular performance with advancing age would be associated, in varying degrees in different individuals, with the changes listed above, i.e., increases in vascular stiffness; increases in systolic, pulse, and mean arterial pressure; decreases in left ventricular compliance and contractility; lesser myocardial responsiveness to β sympathetic agonists; and declining efficiency of cellular enzyme systems. However, some investigators who have studied longitudinally selected populations believe that many of these disabilities are due to associated pathology, especially coronary artery disease, and not to aging per se.[2,6] Certainly the separation of changes due to disease and those due to aging is often very difficult.

II. ORTHOSTATIC CARDIOVASCULAR RESPONSE

A. IMMEDIATE HEMODYNAMIC RESPONSE

Free-stand or head-up tilt induces a rapid, immediate HR increase, the nature and degree of which are used, particularly by European investigators, as an index of autonomic neuropathy.[10] The characteristics of this "initial heart rate complex" are influenced by the speed of attaining the head-up posture, the respiratory phase at the time, and other factors, as described in Chapter 1. There is increasing evidence that age importantly affects this response.

Dambrink and Wieling recorded the initial HR response to free stand and head-up tilt in subjects from l0 to 90 years of age.[10] They found a more prominent HR response to free stand than to head-up tilt; with increasing age, there was a lesser HR and diastolic pressure response but no difference in the systolic pressure. The main finding was a progressive decline in the immediate HR rise with age, which they thought might be related to decreased responsiveness of the arterial baroreceptor reflex and/or to an age-related decline in the circulatory response to adrenergic stimuli.[10]

The peak of the immediate HR response occurs in about 12 to 16 s, followed by a fall to a low point in about 25 to 32 s. Some investigators have quantitated the initial HR complex on the basis of the R-R interval ratio at 30 s and at 15 s, i.e., the so-called 30:15 ratio. Gautschy et al., after comparison of several autonomic function tests, found the 30:15 ratio and the beat-to-beat HR variation during deep breathing to be the best indices of parasympathetic cardiac integrity. A well-functioning cardiac vagal reflex with a high initial (15 s) HR peak and a low HR rebound (at 30 s) would result in a relatively high 30:15 R-R interval ratio. The 30:15 ratio decreases progressively with age from about 1.07 in 20 to 39 year olds, to 0.99 to 1.03 in 70 to 92 year olds.[11] Vita et al. found good correlation between the 30:15 ratio and other parasympathetic tests in different age groups.[12] Clark and Mapstone tested the ratio in a wide range of normal subjects, 31 to 92 years of age, during free stand and defined the lower tolerance limit (i.e., the lower fifth percentile) of the 30:15 R-R ratio as 1.01 at 65 years of age.[13]

Smith et al. recorded the immediate circulatory responses to rapid (2.2 s) head-up tilt in young (20 to 29 years), middle-aged (40 to 49 years), and older (60 to 69 years) healthy normal males. The HR rise began within 2 s, peaked at about 8 to 14 s (depending on age), and then began a slow recession. The initial HR rise was less and reached its peak more slowly in the older men (Figure 2).[14] There was a rapid decline in thoracic blood volume (i.e., increase in transthoracic Z_0) in all three groups, but a tendency for a faster and greater decline in thoracic blood volume (TBV) in the two older groups (Figure 2, lower left). Stroke volumes decreased progressively with no significant differences between the age groups.[14]

Cardiac outputs showed an initial surge above control levels during the first 10 to 20 s (due to the sudden increase in HR), after which there began a slow decline (Figure 2, lower right). These findings suggest that the immediate HR rise, by bolstering the cardiac output, may play an important protective role in maintaining arterial pressure and cerebral perfusion during the critical first 5 to 15 s of the head-up posture. It may be that in certain cases of postural intolerance, e.g., those occurring in the very old or in endurance athletes, a deficient cardioacceleration may be responsible for postural hypotension.[16]

It is interesting that the greatest HR differential in the rapid head-up tilt was between the oldest and two younger groups, but the greatest differential in TBV decrease was between the youngest and the two older groups (Figure 2). During this study of "transient" responses to head-up tilt, it was also noted that upon quick return of the subjects to horizontal, there was a marked, rapid overshoot of HR and TBV of about 3-to 5 s duration. The rebound of stroke volume and cardiac output was slower and less marked. As a part of this investigation,[14] second-by-second hemodynamic values were also determined during the Valsalva maneuver in the same subject groups. Curiously enough, in contrast to the findings in rapid head-up tilt, the decrease in TBV during the 40-mmHg Valsalva was *least* in the oldest group.

B. STABILIZED HEMODYNAMIC RESPONSE
1. Heart Rate
In the following will be considered hemodynamic changes occurring after about 5 to 10 min of adaptation to the erect posture or lower body negative pressure (LBNP). These changes are quite different from the immediate responses previously considered. Compared with young adults, HR increments between the supine and erect position were much reduced in subjects aged 40 to 56 years,[15,16,19] and in subjects aged 75 to 86 years, sometimes to less than one half.[17,18] There were also much lesser degrees of tachycardia at –40 mmHg LBNP in older adults[20-22] and lesser increases in HR for a given fall in systolic pressure.[17] However, Nixon et al. reported no significant change in HR in either young or old subjects during LBNP.[21]

2. Arterial Pressure and Vascular Resistance
Several investigators have reported higher supine systolic, diastolic, and mean arterial pressures[1,3,6,16,21] and lesser increases in diastolic and mean arterial pressures in older subjects in response to the head-up position.[1,15,16] Most investigators reported no significant changes in systolic pressures during standing or head-up tilt[15,16,19] and no significant changes in arterial pressures during –40 mmHg LBNP.[20-22] However, some observers have noted considerable instability of arterial pressure in healthy subjects 75 years of age and older upon standing; about 20% of geriatric unit residents experienced decreases of greater than or equal to 20 mmHg of systolic pressure along with symptoms of dizziness within 2 min of standing.[17]

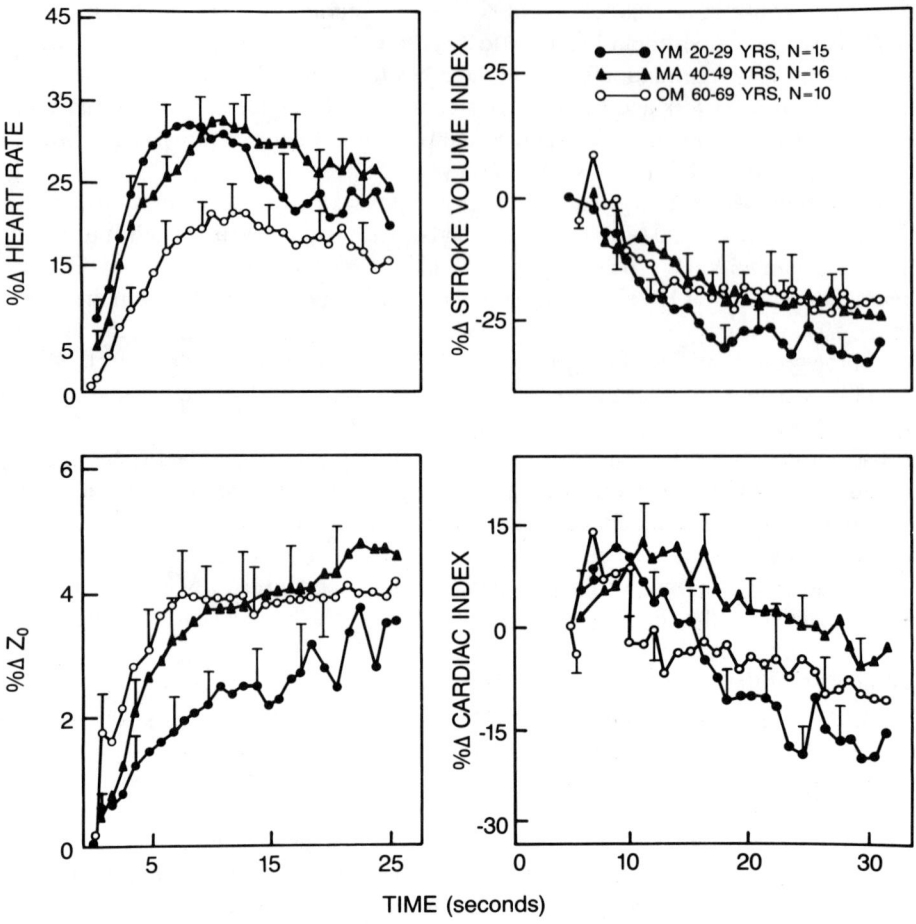

FIGURE 2. Mean percent change in heart rate (upper left), transthoracic impedance (lower left), stroke volume index (upper right), and cardiac index (lower right), during first 30 s following rapid shift to 70° head-up tilt in healthy males of different ages. Note the delay and reduction in heart rate peak in oldest group (upper left) and a tendency (non-significant) toward thoracic blood volume decreases (i.e., greater increases in % Z_0) in the two older groups (lower left). (From Smith J. J., Barney J. A., Porth C. J., Groban, L., Stadnicka, A., and Ebert, T. J., *Physiologist,* 27, 210, 1984. With permission.)

An increased resting total peripheral resistance (TPR) is a common finding in older healthy subjects,[1,3,6,16,19,20] but as previously noted, in some older individuals in whom systolic and diastolic pressures are not elevated, there may not be an increased TPR.[6] In some studies, TPR increases were greater in older subjects during LBNP[20,22] and freestand,[19] but in two head-up tilt studies there were no significant age differences in TPR response.[16,25] The possible mechanisms of the aging changes in vascular resistance are discussed in a following section.

3. Stroke Volume and Cardiac Output

Several investigators studied the decreases in stroke volume and cardiac output that are characteristic of orthostatic stress; in two LBNP studies, the decreases were 20 to 40% less in older than in younger subjects.[20,22] In 70° head-up tilt, one group found lesser supine values of stroke volume and cardiac output in older subjects, but no age differences in the changes with tilt;[16] a second group reported smaller supine stroke volumes in the elderly and lesser changes during tilt;[23] and a third group, using pulsed-Doppler techniques, did not report supine values but found distinctly lesser percent decreases in stroke volume and cardiac output in older subjects, 1 min after head-up tilt.[24]

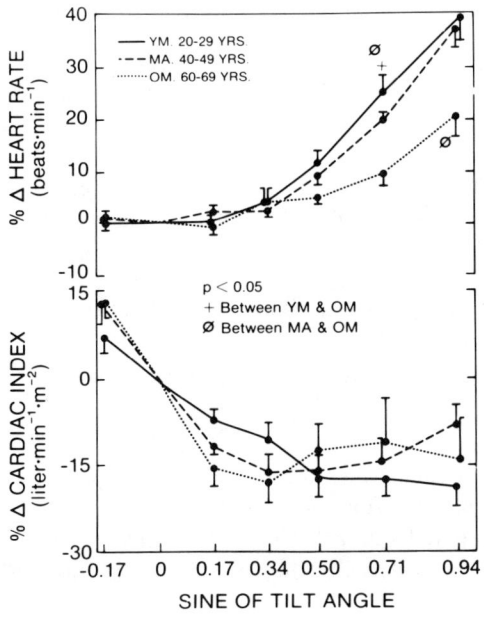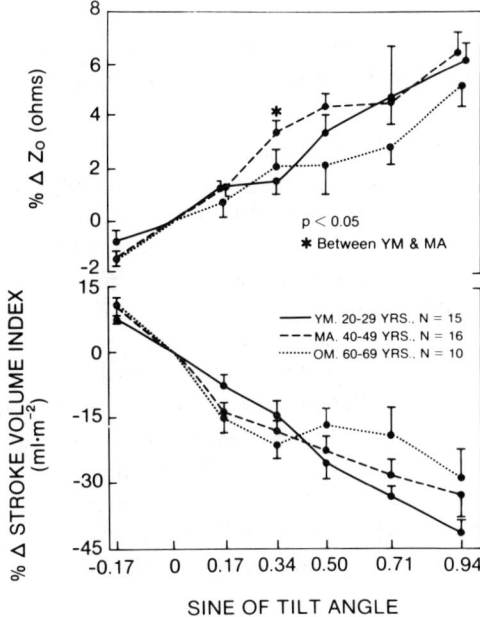

FIGURE 3. Mean percent change in heart rate (upper left), cardiac index (lower left), transthoracic baseline impedance (Z_0) (upper right), and stroke volume index (lower right) at different head-up and at $10°$ head-down tilt angles in young (YM), middle-aged (MA), and older (OM) men. The sine of the angle represent the gravity vector and is a better index of the degree of stress than the angle itself. (From Smith, J. J., Hughes, C. V., Ptacin, M. J., Barney, J. A., Tristani, F. E., and Ebert, T. J., *J. Gerontol.*, 42, 406, 1987. With permission.)

In a graded head-up tilt study, the stroke volume decreases in young, middle-aged, and older males were progressive and had an approximately linear relationship (especially in the younger group) with increasing gravity increments (Figure 3, right lower). There were no significant age differences in these stroke volume changes, and it was evident that the decreases in stroke volume were heavily gravity dependent. The decreases in TBV (increases in transthoracic Zo) were also related to increasing gravity increments in near-linear fashion and, therefore, also to the decreases in SVI (Figure 3, right upper and lower), suggesting that, under these experimental conditions, in healthy males up to about age 70, there is a good consistency of the Frank-Starling response through a wide spectrum of different preload stresses.[16]

In this same study, the pattern of cardiac output decrease was quite different from that of the stroke volume decrease; the cardiac outputs decreased promptly at lower tilt angles (10 to 20%), then plateaued and remained essentially unchanged at the higher tilt angles (Figure 3, left lower). In the face of the marked decreases of SVI in the head-up position (25 to 40%), it would appear that the HR increases are important in maintaining cardiac output in the stabilized, as well as in the immediate response to the erect posture.[16]

4. Thoracic Blood Volume (TBV)

Because of the paucity of available methods, very few studies of TBV have been done. (Methods for determining thoracic [or central] blood volume are discussed in Chapter 1.) There is suggestive evidence that supine TBV is less in older than younger subjects.[11,14,18] There appear to be age-related differences in TBV changes during orthostasis, depending to some extent on the type of orthostatic stress and the sex of the subject. In male subjects at −40 mHg LBNP, there was a lesser decrease in TBV in older compared with younger men.[20] However, during head-up tilt in male subjects, there were no significant age differences in TBV responses,[16] and in free-stand experiments involving an equal mixture of male and female subjects, there was an average of 45% *greater* decrease in TBV in older compared with younger subjects.[19] Sex has here been specified because, as discussed in Chapter 5, several hemodynamic changes during orthostatic and other stresses are importantly influenced by sex. Frey and Hoffler, using the impedance method, found greater decreases in TBV in women than men during LBNP.[22]

5. Cardiac Function

Nixon et al. used two-dimensional echocardiography to study left-ventricular function in older subjects; changes in left-ventricular end-diastolic volume and stroke volume in response to –40 mmHg LBNP were similar, but of smaller magnitude, in older subjects.[21] End-systolic volume (EDV) did not change during LBNP, but control EDVs were somewhat greater in older subjects; the results suggested an age-related increase in diastolic stiffness,[21] which is in agreement with previous findings.[6]

Reports by Frey et al. indicated that during head-up tilt the mean stroke ejection rate[19] and during LBNP the Heather index[22] (a measure of ventricular contractility), were both decreased in older subjects compared with younger ones.

Using radionuclide angiography, Port et al. found no age effect at rest on left ventricular ejection fraction (LVEF), EDV, or regional wall motion, but during exercise the LVEF was less than 0.60 in 45% of subjects over 60 years of age, compared with a 2% incidence in a younger control group; furthermore, in the older subjects there was an increased incidence of wall motion abnormality.[9]

The effects of age on myocardial function are difficult to determine for two reasons: ventricular "contractility" can theoretically only be determined by controlling preload, after-load, and HR and then applying a series of standard ventricular stimuli; these conditions, difficult to produce in animal experiments, are virtually impossible to achieve in the human. In addition, as previously mentioned, the separation of a chronic pathological effect, especially coronary disease, from true aging is a formidable task.[2]

III. MECHANISM OF AGING RESPONSE TO ORTHOSTASIS

A. GENERAL

The alteration in the response of older individuals to orthostatic stress is due mainly to two factors: changes in autonomic function, and the structural and functional changes in the circulatory system itself. Some of the latter factors have been previously mentioned, e.g., the decrease in arterial distensibility, ventricular compliance, and ventricular contractile ability along with an increase in total vascular resistance and a consequent heightened cardiac afterload. It is likely that in most subjects both autonomic and physical factors are responsible for the aging changes.

B. BAROREFLEXES AND AGING

Many studies have indicated that responses to different stresses, including head-up tilt and LBNP, are reduced with advancing age.[10-20] On the presumption that these altered responses might involve deficiencies of reflex function, investigators have determined baroreflex sensitivity by calculating the HR-systolic pressure slope after a drug-induced increase of arterial pressure. Such studies have shown a negative correlation between age and baroreflex sensitivity.[27,28] In one study, there was no correlation between age and resting arterial pressure, but increasing arterial pressure (as well as age) acted independently to reduce baroreflex sensitivity.[28] Later reports of immediate HR responses to sudden increments of neck suction and neck pressure indicated a lesser tachycardia during decreased carotid pressure and lesser bradycardia to increased pressure with age in both women[30] and men.[31] Most early studies of the age effect on baroreflexes involved an analysis of cardiac rather than peripheral vascular responses.

C. GENERAL AUTONOMIC MECHANISMS IN THE ELDERLY

1. Cardiac Vagal Responses

As determined by the respiratory sinus arrythmia method, there is, with advancing age in human subjects, a decreased resting cardiac vagal tone in both men[31,34] and women.[30] The tachycardic responses to carotid sinus stimulation,[30,31] to rapid head-up tilt,[14] and to free stand,[10,34] as well as the bradycardic response to face immersion,[35] are all blunted in older subjects due to a vagal influence. As discussed in Chapter 1, autonomic blockade experiments have demonstrated that immediate (first 2 to 5 s) HR changes after stresses such as head-up posture are under vagal control, but in subsequent seconds a combined vagal and sympathetic cardiac influence is manifest.[34,54,56,57] In the longer term, i.e., upon stabilization following experimental alterations in arterial pressure,[27,28] head-up tilt,[11,15,16] or LBNP,[20,22] HR responses are also diminished in older individuals. Since the heart is then under a combined vagal-sympathetic influence, the exact mechanism of the decreased HR response in such cases is not certain.[57]

2. Cardiac Sympathetic Responses

Lakatta and others have demonstrated a decreased myocardial responsiveness to isoproterenol and other adrenergic agents in both aged animals and humans.[6] Vestal et al. found that there was, with advancing age in healthy males, a diminished responsiveness to both sympathetic agonists and antagonists.[38] Roberts and Steinberg reported an inverse correlation of age with adrenergic control of cardiac function.[39] It was suggested that this insensitivity might be due to a deficiency of central autonomic control, to a decreased density of effector cells,[39] or to the inability of aged tissue to develop a supersensitivity response.[40] While there have been some conflicting reports, the prevailing opinion is that in most cases the responses of cardiac β receptors are either reduced or unchanged,[41,42] and that the reduced uptake of the cardiac receptors may play a role in potentiating the heightened norepinephrine (NE) stress response characteristic of the elderly.[41]

In a study of orthostatic stress, DeCaprio et al. found that in 45 healthy individuals, 15 to 82 years, the QT/QS_2 ratio increased with standing and that the changes were significantly correlated with age. Propranolol prevented the increase. Since the QT/QS_2 ratio is an index of the effect of β adrenergic stimulation, the results suggested a lessened cardiac rate response to sympathetic stimulation.[26]

D. ADRENERGIC RESPONSES AND AGING

It has been shown that the resting levels of plasma NE,[26,36,41] as well as NE increases in the head-up posture,[29,36] are greater in older subjects. Sowers et al. studied NE responses of healthy young, middle-aged, and older men, all of whom were within 5% of their ideal body weight.[36] Their basal plasma NE concentrations were positively correlated with age, and the increases in plasma NE concentration, 10 min after assumption of the head-up posture, were also directly related to age. The Δ NE values were 70% greater in the middle-aged and 27% greater in the older subjects than in the younger[36] (Figure 4). Thus, the effect of aging on incremental adrenergic activity was independent of adiposity. While plasma NE concentrations were greater during isometric handgrip, the increases were not significantly related to age.

Pratley et al. found that the usual high plasma NE levels at supine rest and the head-up position in the elderly could be significantly reduced by weight loss and exercise training;[37] these authors concluded that the elevated NE levels in older individuals may be due, in part, to obesity and physical inactivity and thus are potentially reversible. Because of the evident blunting of vagal cardiac reflexes and hyperadrenergic responses with advancing age, Frey et al. have suggested that there is a basic autonomic shift with age to a lesser vagal and an increased sympathetic influence.[19]

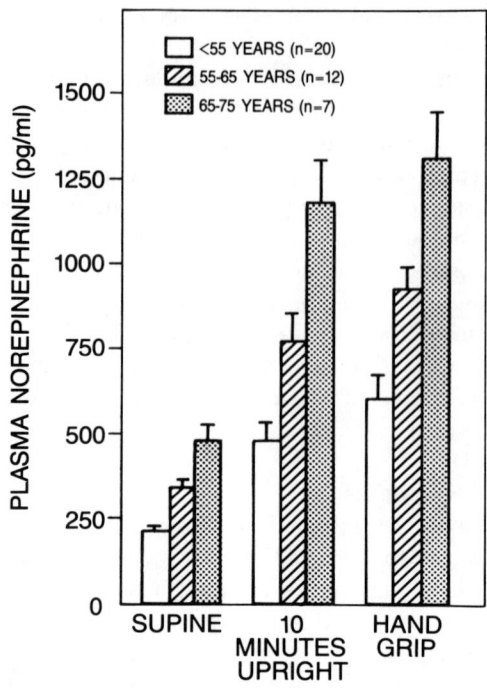

FIGURE 4. Mean (± SEM) plasma norepinephrine responses to 10 min upright posture and 5 min isometric handgrip in three age groups of men. (From Sowers, J. R., Rubenstein, L. Z., and Stern, N., *J. Gerontol.*, 38, 315, 1983. With permission.)

E. VASCULAR RESISTANCE RESPONSES IN THE ELDERLY

The lack of effect of cardiac sympathetic and vagal blockade on general orthostatic tolerance (Chapter 1) strongly suggests that, in the absence of cardiopathology, orthostatic tolerance is more dependent on TPR than on cardiac performance. Yet compared with cardiac reflexes, relatively few studies have been done on reflex control of TPR during postural change, and fewer yet on how aging affects this response. As mentioned in Section II above, resting TPR generally increases with age,[1,3,6,16,19] but during orthostasis the result apparently depends on the type of stress applied. TPR responses are greater in older than younger subjects during free stand[19] and LBNP,[20,22] but apparently are similar during head-up tilt.[16,25] However, Mohanty et al. reported that low-level LBNP resulted in lesser increases in forearm vascular resistance in older subjects,[20] suggesting an age-related deficiency in cardiopulmonary baroreflex function.[29] Kaijser and Sachs stated that the usual hypertensive response to isometric handgrip was attentuated with age.[35]

While there is no unanimity on the effect of aging on the vascular resistance response to orthostasis, the few studies done thus far suggest a general tendency toward either maintenance or augmentation of TPR during orthostatic stress in (mostly male) subjects up to the ages of 60 to 70 years. This result might theoretically be due to the hyperadrenergic state in the elderly or the generalized arteriosclerosis that is characteristic of the aged. While elevated plasma NE levels undoubtedly exist in older subjects, the relatively minor hemodynamic effect of circulating catecholamines in comparison with the sympathetic neurogenic influence would seem to make this a less likely possibility. The studies of Folkow and others have provided evidence that increased vessel-wall thickness, such as commonly occurs with aging, may predispose to disproportionate vascular resistance responses during stress because of the fourth power effect on the radius in the Poiseuille equation.[58] In our opinion, the latter is the more likely, though still unproven, possibility.

There is increasing evidence that in the human the cardiopulmonary baroreceptors play a major role in mobilizing vascular resistance in orthostatic stress. At low levels of LBNP[20,55] and head-up tilt,[16] at which point there are no significant changes in arterial pressure or HR, increases in TPR occur that equal 70% or more of the total TPR achieved at full orthostatic stress of –40 mm Hg LBNP and 70° head-up tilt. Thus, the bulk of the total vascular resistance response is mobilized at low-level orthostatic stress, at which time it is likely that only cardiopulmonary baroreflexes are active. At full orthostatic stress, it is assumed that both cardiopulmonary and arterial baroreflexes are operative.

Using the microneurographic method in subjects 23 to 56 years of age, Burke et al. found almost uniformly an increase in the burst incidence in muscle sympathetic nerve activity (MSNA) upon a change from the supine to the sitting position.[32] With similar methods, Sundloff and Wallin found marked interindividual differences in MSNA burst incidence in the limbs of a human subject; there were no interindividual correlations of burst incidence with arterial pressure, but a slight tendency for the incidence to increase with age.[33] This increased MSNA tendency with age may be related to the prevailing tendency toward TPR maintenance or augmentation with age. However, as discussed in Chapter 1, there may not always be a close parallelism of MSNA burst frequency and peripheral vascular resistance.

This disassociation of a sympathetic vasoconstrictor tendency and the peripheral resistance may be partly due to the vagaries inherent in the calculation of vascular resistance as the quotient of pressure difference and tissue flow. This simplified hemodynamic equation may not be representative of the degree of sympathetic vasoconstriction, because of autoregulation and other local adjustments of arterial and venous tone, particularily if the flow rate across the vascular bed is changing, as it usually is during orthostasis. However, microneurographic MSNA determinations in a limb must certainly be a valid indication of basic sympathetic vasoconstrictor tendencies in limb skeletal muscle (Chapter 1).

F. ORTHOSTATIC TOLERANCE IN THE ELDERLY

The characteristics, diagnosis, and treatment of the specific disease entity known as orthostatic hypotension are discussed in Chapter 6. In the following, we will consider a more general type of orthostatic intolerance that develops in older people and in which the disability is believed to be due primarily to the circulatory and neurophysiological changes that occur with age. While orthostatic tolerance is usually quite well maintained in healthy elderly to about 60 to 70 years,[16] after 70 there is an increasing tendency toward an instability of arterial pressure and toward postural hypotension, i.e., a decrease of ≥ 20 mmHg in systolic or diastolic pressure upon standing.[17] The precise cause of this phenomenon is not presently known. While orthostatic hypotension may be both diastolic and systolic, that associated with age seems to be more often a systolic hypotension. A failure of arterial pressure may, of course, be due to a deficiency of cardiac output or total peripheral resistance (TPR). Cardiac output in the erect position, is, in turn, mainly dependent on HR, left ventricular ejection, and the maintenance of venous return through the action of autonomic reflexes and the skeletal muscle pump (SMP), as described in Chapter 1, Section VI.

From the standpoint of the hemodynamic changes that usually occur with age, as described in Sections I and II above, there are five main deficiencies in the circulatory response to orthostasis that might theoretically predispose to orthostatic hypotension. These are defects in cardioacceleration, ventricular ejection, TPR, SMP, and thoracic blood volume (TBV).

As described in the previous sections, there is clear evidence that in many elderly persons, there is a diminished capacity for cardioacceleration and for the usual increase in ventricular contractility during stress. While previous studies in younger subjects with cardioblocking agents indicate that in the absence of cardiac pathology cardiac response is generally less important in the determination of orthostatic tolerance than is TPR, this may not hold in elderly subjects in whom there may be serious functional defects in cardiac response. It should also be pointed out that, aside from the direct effect on cardiac output, a diminished cardioacceleration might also influence vascular resistance, particularly in the skeletal muscle beds, through the HR-vascular resistance link described by Wallin and his colleagues.[32,33] The possible role of a deficient cardioacceleration and inadequate left ventricular contraction should be further studied in orthostatic intolerance of the elderly.

Decreased blood volume is recognized as a common cause of general orthostatic intolerance, particularly of the acute type. There is relatively little data on the age effect on total or thoracic blood volume, but there is indirect evidence that supine TBV is less in older than in younger subjects. As described in the following section, plasma volume depletion is undoubtedly a factor in acute orthostatic tolerance in the elderly.[53] The role of total or thoracic blood volume in chronic orthostatic hypotension of the elderly is unknown.

On the basis of hemodynamic principles and previous cardioblocking experiments, TPR would loom as a key factor affecting arterial blood pressure during orthostasis. Previous studies seem to indicate a maintenance or augmentation of TPR during orthostasis in the elderly.[16,19,20,22,25] However, we believe this question is still unsettled for several reasons: first, most previous studies have involved subjects under 70 years of age and orthostatic hypotension becomes more common at age 70 and above. Furthermore, as previously mentioned, the numerical calculation of TPR changes on the basis of the abbreviated Poiseuille equation may be misleading as an index of vasoconstrictor tendency. Moreover, in the elderly the resting TPR is frequently increased due to generalized arteriosclerosis; this considerably complicates the comparative significance of increases of TPR. We believe, therefore, that further careful studies of TPR changes, as well as of regional resistance changes, should be carried out in subjects of 70 years or older. There should also be further investigation of the relative roles of the cardiopulmonary and arterial baroreceptor reflexes in the control of vascular resistance in the elderly.

Finally, as discussed in detail in Chapter 1, Section VI.E, a number of cardiovascular physiologists believe that the SMP plays a critical role in assisting venous return in the upright position and that the importance of the SMP in orthostatic tolerance has been considerably underestimated. In most elderly, decreased physical activity undoubtedly leads to various degrees of deconditioning of the antigravity muscles of the lower body and, most likely, diminished SMP function. We believe that the possible role of SMP hypofunction in orthostatic intolerance of the elderly would definitely warrant further investigation.

IV. FACTORS AFFECTING POSTURAL TOLERANCE IN THE ELDERLY

Since this subject has been discussed at some length in Chapter 1, only those aspects relevant to age will be considered here. While factors affecting orthostatic tolerance have been studied extensively in young subjects, relatively few such studies have been made in the elderly. In one of these, Lipsitz et al., in a comparative analysis, found that the fainting rate in vagally mediated syncope was less in older ($\bar{x} = 81$ years) than in younger ($\bar{x} = 21$ years) subjects, but in both groups β blockade was ineffective in reducing the fainting rate.[43]

A. SYSTEMIC DISEASE

Drugs, dysautonomia, and most systemic diseases will generally decrease resistance to all physiological stresses, including orthostasis. One of the most common ailments of the elderly, which may particularly influence orthostatic tolerance, is diabetes; this disease affects 6% of all adults in the U.S. and Europe, as well as 16% of the population over 65 years. It is the most common cause of autonomic neuropathy, which, in turn, is a primary cause of postural intolerance.[13]

B. AMBIENT TEMPERATURE

Increased external temperature has long been known to decrease orthostatic tolerance in the young and the old; defects in thermal regulation would be expected to intensify the problem. Shiraki et al. studied head-up tolerance in young and old subjects at relatively normal ambient temperatures and after 105 min of exposure to 40°C and 40% relative humidity.[44] These investigators found that cardioacceleration during the tilt test was less in the older men and forearm blood flow responses were altered, suggesting deterioration of sympathetic reflexes in the aged; however, overall orthostatic tolerance to tilt was not significantly different in the two groups.[44] Collins et al. noted that the elderly had poor vasoconstrictor response to cooling, indicating a definite deficiency in thermal capacity.[17] In diabetics, this situation is intensified, since in this condition A-V O_2 differences are low, with a resultant critical depression of metabolic heat production due to failure of O_2 utilization.[17]

C. INITIAL HEMODYNAMIC VALUES

As described in Chapter 1, according to the law of initial values, the degree of individual hemodynamic response to stress is often negatively correlated with the control level of that variable.[45] Burke et al. found that the changes in muscle sympathetic nerve burst frequency in the human limb during postural change was inversely related to the resting level of nerve activity in young males.[32] Smith et al. noted in young males a high positive correlation between the supine values of HR and, diastolic, pulse, and mean arterial pressure and their respective values in the sitting and standing position; there were also high negative correlations between the supine values of systolic and pulse pressure and the delta values in head-up tilt and standing.[46] Kalbfleisch et al. similarily observed, in both young and older male subjects, a positive correlation of control values of HR and arterial pressures with values at 70° head-up tilt and a prevailing tendency toward negative correlation between the control values and the increments between control and tilt.[15] MacLennan et al. also reported in older subjects, correlations between resting and delta values of arterial pressure during postural change.[47]

Harris et al. noted that a 20-mmHg orthostatic drop in systolic blood pressure was not independently associated with age, but increased in incidence with increasing supine systolic pressure; such "positional drop" in systolic pressure (20 mmHg or more) occurred in 5% of those with systolic pressures of 120 to 139 mmHg, but in 28% of those with systolic pressures greater than or equal to 160 mmHg; these authors concluded that the increase in postural hypotension with age was related to the increased prevalence of systolic hypertension with age, rather than with age per se.[48]

D. POSTPRANDIAL HYPOTENSION

Although pressure fall after meals had been previously reported in frail, institutionalized subjects, Lipsitz and Fullerton found, in healthy elderly ($\bar{x}$ = 73 years), a fall in systolic pressure ($\bar{x}$ = 11 mm Hg) and a rise in HR ($\bar{x}$ = 7/min) during a 2-min free stand, 60 min after breakfast.[49] There was a highly significant inverse correlation between the basal sitting systolic pressure and the change in systolic pressure from sitting to standing. These subjects did not have postural hypotension. Since resting blood pressure is usually negatively correlated with the degree of postprandial hypotension, these authors raised the interesting question of whether the postprandial hypotension should not be considered in the diagnosis and treatment of hypertension.[49]

Westenend et al. reported a maximum decrease in systolic pressure (7 mmHg) and, diastolic pressure (14 mmHg) and an increase in HR (5 bpm) in elderly subjects ($\bar{x}$= 79 years) about 50 min after breakfast. In a younger group, blood pressure did not change. There was a significant correlation between preprandial mean arterial pressure (MAP) and the MAP decrease during free stand after breakfast.[50] Kelbaek et al.[51] reported in young, healthy subjects (19 to 29 years), a 61% increase in cardiac output (due mainly to stroke volume increase) and a 17% increase in HR after a standard meal; there were no changes in MAP but a 29% increase in left ventricular

EDV. Since there was no change in blood pressure, it was evident that the TPR decreased markedly after the meal. The stroke volume and cardiac output changes were dependent on the autonomic nervous system, since they did not occur after autonomic blockade (metropolol and atropine). These findings are consistent with a postprandial splanchnic vasodilation.[51]

Onrot et al. found substantial blood pressure decreases (28/18 mmHg) in patients with autonomic failure 60 min after a standard meal. Long-term administration of caffeine significantly reduced the degree of postprandial hypotension in these patients, but they believed that this effect was not due to a sympathoadrenal or renin-angiotensin effect; they cited reports of eventual tolerance to this pressor effect of caffeine.[52]

While the mechanism of postprandial hypotension in the elderly is unknown, two theories have been advanced, i.e., sequestration of splanchnic blood volume[49,50] or altered insulin release in the elderly.[50] Postprandial hypotension is most marked after breakfast.[52] Since the hepato-splanchnic portion of cardiac output is large (25%) and its vascular capacity great, it would be reasonable to assume that increased metabolic demands of the splanchnic organs incident to digestion might temporarily sequester a sizeable portion of the blood volume, which might then be denied to the circulation during the head-up posture. It might also be theorized that age-related failure of insulin-mediated sympathetic nervous system activation[49] or the release of other hormones (e.g., adenosine[52]) may, in concert with a depressed cardio-accelerator reflex and diminished SMP capacity in the elderly, further reduce the ability of the circulatory system to combat the splanchnic blood sequestration.

E. SODIUM AND WATER DEPLETION

The depletion of sodium and water probably constitutes a very real handicap for successful adaptation to almost any circulatory stress; this possibility was investigated in an interesting study by Shannon et al.[53] They determined the hemodynamic response to a 3-min, 60° head-up tilt before and after diuretic-induced Na depletion in healthy young (25 to 35 years) and older (65 to 80 years) subjects. Younger subjects maintained their systolic pressure before and after diuresis in response to orthostasis; older subjects maintained systolic pressure before diuresis, but afterwards experienced an average 22 mmHg decline during tilt (Figure 5). Older subjects also had lesser increases in HR and three of six older subjects had postural symptoms. Diastolic pressure did not fall during tilt in older subjects, suggesting that there may not have been a vasoconstrictor failure. It was interesting that the rate of cardioacceleration was greater in the first 18 s after diuresis than before, the increase being more pronounced in the young.[53]

These results suggest that healthy elderly may be able to compensate for the caudal sequestration of blood incident to the head-up position, but that even a mild Na and H_2O depletion may make the older subject more vulnerable to a hypotensive episode during orthostasis.

V. SUMMARY AND CONCLUSIONS

A. GENERAL CIRCULATORY CHANGES WITH AGE

With increasing age there is a progressive decline in resting cardiac output, i.e., about a 37% decrease between age 30 and age 80.[5] This is mainly due to a lesser metabolic rate, which in turn is probably the result of a decreasing efficiency of the tissue oxidative enzymes. In most elderly there is a progressive, generalized arteriosclerosis, lesser arterial distensibility, slower relaxation and contraction of the ventricles, and steady increases in arterial pressure and total vascular resistance. In healthy elderly, these changes impose relatively little handicap at rest. However, circulatory stresses such as exercise uncover the erosion of the physiological reserve. Maximum work capacity (Vo_{2max}) decreases about 1%/year, which means about a 40% decline between ages 25 and 65[8] (Figure 1).

These changes, however, are subject to large individual differences, depending upon physical

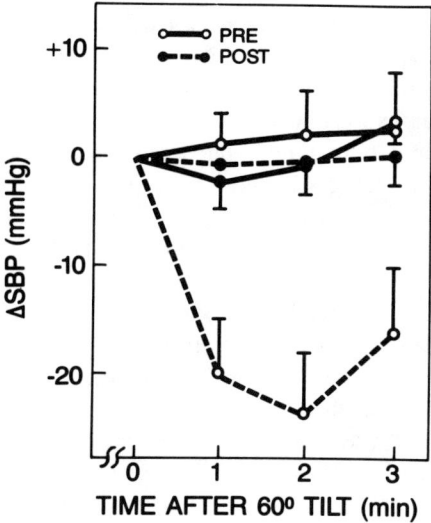

FIGURE 5. Systolic blood pressure (SBP) responses to 3 min 60° upright tilt in healthy, young, 25 to 35 year (dark circles) and healthy, older, 65 to 80 year (light circle) subjects before and after a moderate 2-d diuretic-induced Na depletion. Three of the six older subjects had symptoms of postural intolerance. (From Shannon, R. P., Wei, J. Y., Rosa, R. M., et al., *Hypertension,* 8, 438, 1986. With permission.)

fitness as well as nutritional, socioeconomic, genetic, and other factors. In certain select groups, e.g., in the Baltimore Longitudinal Study, investigators have found that circulatory function is much better preserved and have suggested that many of the reported circulatory deficiencies are not due to aging per se but to associated disease processes, especially coronary heart disease.[2,6]

B. AGE AND THE ORTHOSTATIC RESPONSE

The *immediate response* to free stand or rapid head-up tilt is a rapid 30 to 40% rise in HR in the young adult; this increase begins within 2 s, peaks in about 12 to 16 s, and then usually decreases to a lower value at or about 25 to 30 s. With increasing age there is a decline in the peak HR and a tendency toward a greater fall in thoracic blood volume (TBV)[14] (Figure 2). This initial heart rate complex is used clinically as an index of autonomic neuropathy.[10,11,34,54]

During the first 10 to 15 s after tilt, cardiac output shows an initial surge above control levels due to the sudden increase in HR (Figure 2); this suggests that this immediate HR rise may be instrumental in maintaining arterial pressure and cerebral perfusion during the critical first seconds of the upright posture.[14] A diminished cardioacceleration in the elderly and the endurance athlete may play a role in the postural intolerance reported in these subjects.

In the *Stabilized hemodynamic response,* i.e., at 5 to 10 min after upright posture or lower body negative pressure (LBNP), there are lesser increases in HR and diastolic and mean arterial pressure in older subjects compared with younger ones. [1,3,6,15,16,21] After age 70, there is often instability of arterial pressure after postural change and an increasing tendency toward postural hypotension, i.e., a fall of greater than or equal to 20-mm Hg of systolic pressure upon standing.[17]

In both LBNP and the head-up position, there are progressive decreases in stroke volume and cardiac output; in two LBNP studies, the decreases were 20 to 40% less in older than in younger subjects,[20,22] but in a study of head-up tilt there were no significant age differences in these two variables[16]. In one graded, head-up-tilt study, there were near-linear, inverse relationships between increasing gravity increments and decreases in both TBV and stroke volume, indicating that both of these variables were heavily gravity dependent[16] (Figure 3).

An increased total peripheral resistance (TPR) is commonly found in older healthy individuals at rest.[1,3,6,16,19,20] TPR increases were also 40 to 100% greater in older, compared with younger, subjects during LBNP[20,22] and free stand,[19] but in two head-up-tilt studies, there were no significant differences in TPR changes between old and young subjects.[16,25]

C. MECHANISM OF AGING RESPONSE TO ORTHOSTASIS

The alteration of the orthostatic response to aging is usually due either to changes in autonomic function, to circulatory inadequacy, or to both. Studies have clearly indicated a lessening of arterial baroreflex control of the heart in older subjects. This has been manifested by a blunting of reflex effects in both vagal and sympathetic cardiac control mechanisms.[27-31] Experiments in animals and humans have also shown a decreased responsiveness of the myocardium to both sympathetic agonists and antagonists.[6,38-41]

Resting levels of plasma NE,[2,6,36,41] as well as NE increases in the head-up posture,[29,36] are greater in older than in younger subjects (Figure 4), so that a hyperadrenergic state undoubtedly exists in the elderly subject. It is believed that this is probably linked in a cause-effect manner with the depressed sensitivity of cardiac β receptors to sympathetic agonists.[41] Recent studies have suggested that the high plasma NE levels can be reduced by weight loss and exercise training.[37]

There is increasing evidence that in the human the cardiopulmonary baroreceptors play a major role in mobilizing vascular resistance in orthostatic stress.[16,20,51] There is a general tendency toward increased TPR responses during orthostatic stress in the elderly, but the evidence is conflicting and results seem to depend on the type of stress used and the sex of the subject.[16,19,20,22,25] The role of the cardiopulmonary baroreceptors, if any, on this alteration in vascular resistance is not known.

While orthostatic tolerance is quite well maintained in the elderly until about 70 years of age, beginning at about age 75 there is a progressive tendency toward orthostatic hypotension. The cause of this is presently unknown.[16-18] There are at least five circulatory deficiencies that might theoretically be involved in this predisposition of older subjects toward orthostatic hypotension, e.g., (1) inadequate cardioacceleration in the first 10 to 15 s of orthostasis, resulting in an insufficient cardiac output surge in the early phase of the head-up posture; (2) the slower and lesser response of the myocardium during relaxation and contraction; (3) deconditioning of the antigravity muscles and lesser efficiency of the skeletal muscle pump (SMP); (4) decreased thoracic (and total) blood volume; and (5) inadequate response of TPR to the head-up posture.

While some of these deficiencies undoubtedly exist in older subjects, most have not been systematically studied, so their possible role in the mechanism of postural hypotension is at present only theoretical.

D. FACTORS AFFECTING POSTURAL TOLERANCE IN THE ELDERLY

Aside from the systemic and regulatory influences discussed in Sections I to III, the most important extrinsic factors that reduce postural tolerance in the elderly are

1. *Systemic disease*: particularly cardiovascular disorders and diabetes.[13]
2. *Increased ambient temperature*.[17,44]
3. *Initial hemodynamic values:* there is usually a strong negative correlation between the initial supine value (e.g., blood pressure and HR) and the delta value during change to the upright.[15,32,46,47,48]
4. *Postprandial hypotension*: in subjects over 70 years, a reported 5 to 10% incidence, particularly after breakfast.[49-51] The mechanism is uncertain.
5. *Sodium and water depletion*: a moderate, 2-d, diuretic-induced sodium and water

depletion will cause orthostatic intolerance to head-up tilt in subjects over 65 years of age[53] (Figure 5).

E. CONCLUSIONS AND RECOMMENDATIONS

1. It is believed that it is now clear that age, sex, physical fitness, systemic disorders, and certain environmental conditions are major determinants of orthostatic tolerance and that these parameters should be routinely documented and described in reports on orthostasis.
2. While the responses to free stand, head-up tilt, and LBNP have common elements, there are important differences, particularly in compensatory hemodynamic responses, that should be recognized in assessing orthostatic tolerance. (Chapter 1, Section IX.A)
3. It is believed that consideration should be given to certain investigative approaches, some of which have proven fruitful in the past and some of which have not yet been explored. Among these are: (1) increased application of graded stress procedures, (2) greater use of interindividual data analyses, (3) increased attention to the "immediate" response to orthostatic stress, and (4) assessment of the possible role of the SMP in orthostatic intolerance.
4. Graded stress tends to separate the response elements on the basis of the degree of stress, so that they can be better analyzed and compared; this may furnish clues regarding possible mechanisms.
5. While group averaging is a time-honored and useful technique, inter-individual analysis is often of particular help in uncovering associated factors and mechanisms. The interindividual analyses of Wallin et al.[32,33] and Wolthuis et al.[60] have, e.g., been particularly valuable.
6. The adjustment to the upright position has two phases, the "immediate", i.e., the first 20 to 30 s, and the "delayed", from 2 to 20 min or longer (Chapter 1, Figure 3). Since the occurrence or not of orthostatic intolerance is often established in the first few seconds, it is important to determine hemodynamic changes, including TBV during that period. This might be especially helpful in delineating mechanisms of orthostatic hypotension in the elderly.
7. As emphasized by Ludbrook, Gauer, Rowell, and others, and as described in some detail in Chapter 1, Section VI.E, there is strong, albeit indirect, evidence of the importance of the SMP in maintaining cardiac output in the upright position. In spite of this, very little research effort has been expended in this direction and good methods are lacking. We believe this to be a potentially fruitful line of research, particularly in orthostatic intolerance in the elderly, in whom inactivity and deconditioning of the antigravity muscles may be a common occurrence.
8. Orthostatic hypotension of the elderly is a serious clinical problem. Its importance is due not only to the progressive increase of its incidence with age and its role in the causation of falls in the elderly, but also because it is a major deterrent to ambulation in the aged. Decreased mobility has a crippling effect on the psychosocial activities, independence, and quality of the life of the older citizen. As aptly described by Rowe and Kahn, these elements are of overwhelming importance for successful aging.[59] Orthostatic tolerance is obviously a highly valuable commodity in the aged, but its importance seems in striking contrast to the relative lack of basic cardiovascular investigation being conducted on its mechanisms. We believe serious consideration should be given to increasing the research efforts in this area.

REFERENCES

1. **Strandell, T.,** Circulatory studies in healthy old men, *Acta Med. Scand.*, 175 (414), 1, 1964.
2. **Weisfeldt, M. L., Ed.,** *The Aging Heart,* Raven Press, New York, 1980.
3. **Smith, J. J. and Kampine, J. P.,** Effects of aging, in *Circulatory Physiology — The Essentials,* 2nd ed., Williams & Wilkins, Baltimore, 1984, 237.
4. **Kenney, R. A.,** *Physiology of Aging*: *A Synopsis,* Year Book Medical Publishers, Chicago, 1982.
5. **Guyton, A. C.,** *Cardiac Output & Its Regulation,* W. B. Saunders, Philadelphia, 1963, 9.
6. **Lakatta, E. G.,** Hemodynamic adaptations to stress with advancing age, *Acta Med. Scand.,* Supp. 711, 39, 1986.
7. **Walsh, R. A.,** Cardiovascular effects of the aging process, *Am. J. Med.,* 82 (Suppl. 1B), 34, 1987.
8. **Dehn, M. M. and Bruce, R. A.,** Longitudinal variations in maximal oxygen intake with age and activity, *J. Appl. Physiol.,* 33, 805, 1972.
9. **Port, S., Cobb, F. R., Coleman, R. E., et al.,** Effect of age on the left ventricular ejection fraction in exercise, *N. Engl. J. Med.,* 303, 1133, 1980.
10. **Dambrink, J. H. A. and Wieling, W.,** Circulatory response to postural change in healthy male subjects in relation to age, *Clin. Sci.,* 72, 335, 1987.
11. **Gautschy, B., Weidmann, P., and Gnadinger, M. P.,** Autonomic function tests as related to age and gender in normal man, *Klin. Wochenschr.,* 64, 499, 1986.
12. **Vita, G., Princi, P., Calabro, R., et al.,** Cardiovascular reflex tests, *J. Neurol. Sci.,* 75, 263, 1986.
13. **Clark, C. V. and Mapstone, R.,** Age adjusted tolerance limits for autonomic assessment in the elderly, *Age Ageing,* 15, 221, 1986.
14. **Smith, J. J., Barney, J. A., Porth, C. J., Groban, L., Stadnicka, A., and Ebert, T. J.,** Transient hemodynamic responses to circulatory stress in normal male subjects of different ages, *Physiologist,* 27, 210, 1984.
15. **Kalbfleisch, J. H., Reinke, J. A., Porth, C. J., Smith, J. J., et al.,** Effect of age on circulatory response to postural and Valsalva tests, *Proc. Soc. Exp. Biol. Med.,* 156, 100, 1977.
16. **Smith, J. J., Hughes, C. V., Ptacin, M. J., Barney, J. A., Tristani, F. E., and Ebert, T. J.,** The effects of age on hemodynamic response to graded postural stress in normal men, *J. Geront.,* 42, 406, 1987.
17. **Collins, K. J., Exton-Smith, A. N., James, M. H., and Oliver, D. J.,** Functional changes in autonomic nervous responses with aging, *Age Ageing,* 9, 17, 1980.
18. **Lipsitz, L. A., Maddens, M. E., Pluchino, F. C., Schmitt, W. P., and Wei, J. Y.,** Effect of advanced age on cardiovascular responses to orthostatic stress, *Gerontologist,* 26, 58a, 1986.
19. **Frey, M. A. B., Tomaselli, C. M., and Freeman, M.,** Cardiovascular responses to postural change: differences by sex and age, submitted.
20. **Ebert, T. J., Hughes, C. V., Tristani, F. E., Barney, J. A., and Smith, J. J.,** Effect of age and coronary heart disease on the circulatory responses to graded lower body negative pressure, *Cardiovasc. Res.,* 16, 663, 1982.
21. **Nixon, J. V., Hallmark, H., Page, K., Raven, P. R., and Mitchell, J. H.,** Ventricular performance in human hearts aged 61 to 73 years, *Am. J. Cardiol.,* 56, 932, 1985.
22. **Frey, M. A. B. and Hoffler, G. W.,** Association of sex and age with responses to lower body negative pressure, *J. Appl. Physiol.,* 65, 1752, 1988.
23. **Lye, M. and Vargas, E.,** An analysis of impedance cardiography in the elderly, *J. Med. Eng. Tech.,* 5, 289, 1981.
24. **Tawney, K. W., Johnson, E. C., and Greene, E. R.,** Age-related differences in reactivity of the central circulation to head-up tilt, *Physiologist,* 31, A130, 1988.
25. **Porth, C. M., Barney, J. A., Hughes, C. V., Ptacin, M. J., Sheldahl, L., and Smith, J. J.,** The effect of age and physical fitness on hemodynamic responses to postural stress in women, *Fed. Proc.,* 47, A313, 1988.
26. **DeCaprio, L., Papa, M., Acanfora, et al.,** Effect of aging and beta adrenergic blockage on QT/QS$_2$ changes, *J. Gerontol.,* 44, M3, 1989.
27. **Randall, O., Esler, M., Culp, B., Julius, S., and Zweifler, A.,** Determinants of baroreflex sensitivity in man, *J. Lab. Clin. Med.,* 91, 514, 1978.
28. **Gribbin, B., Pickering, T. G., Sleight, P., and Peto, R.,** Effect of age and high blood pressure on baroreflex sensitivity in man, *Circ. Res.,* 24, 424, 1971.
29. **Mohanty, P. K., Arrowood, J., Sowers, J., McNamara, C., Mulligan, T., and Thames, M. D.,** Effect of age on cardiopulmonary baroreflex, *Circulation,* 4, 350, 1987.
30. **Porth, C. J., Groban, L., and Smith, J. J.,** Carotid-cardiac baroreflex responses decrease with early aging in women, *Physiologist,* 28, 350, 1985.
31. **Smith, J. J., Barney, J. A., Groban, L., et al.,** Carotid-cardiac baroreflex responses decrease with early aging in man, *Fed. Proc.,* 44, 1887, 1985.
32. **Burke, D., Sundlof, G., and Wallin, B. G.,** Postural effects on muscle nerve sympathetic activity in man, *J. Physiol.,* 272, 399, 1977.
33. **Sundlof, E. and Wallin, B. G.,** Human muscle nerve sympathetic activity at rest: relation to blood pressure and age, *J. Physiol.,* 274, 621, 1978.

34. **Karemaker, J. M., Wieling, W. W., and Dunning, A. J.,** Aging and the baroreflex: handbook of hypertension, in *Hypertension in the Elderly*, Vol. 12, Amery, A. and Steussen, J., Eds., Elsevier, New York, 1988.
35. **Kaijser, L. and Sachs, C.,** Autonomic cardiovascular responses in old age, *Clin. Physiol.,* 5, 347, 1985.
36. **Sowers, J. R., Rubenstein, L. Z., and Stern, N.,** Plasma norepinephrine responses to posture and isometric exercise increase with age in the absence of obesity, *J. Gerontol.*, 38, 315, 1983.
37. **Pratley, R., Coon, P., Lumpkin, M., et al.,** Effects of weight loss and exercise training on norepinephrine response to upright posture in older men, *Gerontologist*, 28, 142A, 1988.
38. **Vestal, R. E., Wood, A. J. J., and Shand, D. G.,** Reduced b-adrenoceptor sensitivity in the elderly, *Clin. Pharmacol. Ther.,* 181, 1979.
39. **Roberts, J. and Steinberg, M.,** Effects of aging on adrenergic receptors, *Fed. Proc.*, 45(1), 40, 1986.
40. **Weiss, B., Greenberg, L., and Cantor, E.,** Age-related alterations in the development of adrenergic denervation supersensitivity, *Fed. Proc.*, 38, 1915, 1979.
41. **Docherty, J. R.,** Review, aging and the cardiovascular system, *J. Auton. Pharmacol.*, 6, 77, 1986.
42. **Pfeifer, M. A., Weinberg, C. R., Cook, D., et al.,** Differential changes in autonomic nervous system function with age in man, *Am. J. Med.*, 75, 249, 1983.
43. **Lipsitz, L. A., Clagett, E. R., Koestner, J., et al.,** Reduced susceptibility to vagally-mediated syncope in advanced age, *Physiologist,* 31, A129, 1988.
44. **Shiraki, K., Sagawa, S., Yousef, M. K., et al.,** Physiological responses of aged men to head-up tilt during heat exposure, *J. Appl. Physiol.*, 63, 576, 1987.
45. **Hatch, R. C., Hughes, R. W., and Bozivich, H.,** Effect of resting blood pressure on pressure response to drugs and carotid occlusion, *Am. J. Physiol.,* 213, 1515, 1967.
46. **Smith, J. J., Bonin, M. L., Wiedmeier, V. T., Kalbfleisch, J. H., and McDermott, D. J.,** Cardiorespiratory response of young men to diverse circulatory stresses, *J. Aerospace Med.*, 45, 583, 1974.
47. **MacLennan, W. J., Hall, M. R. P., and Timothy, J. L.,** Postural hypotension in old age. Is it a disorder of the nervous system or of blood vessels? *Age Ageing*, 9, 25, 1980.
48. **Harris, T., Kleinman, J., Lipsitz, L., et al.,** Is age or level of systolic blood pressure related to positional blood pressure change? *Gerontologist,* 26, 59A, 1986.
49. **Lipsitz, L. A. and Fullerton, K. J.,** Postprandial blood pressure reduction in healthy elderly, *J. Am. Geriatr. Soc.*, 34, 267, 1986.
50. **Westenend, M., Lenders, J. W. M., and Thien, T.,** The course of blood pressure after a meal: a difference between young and elderly subjects, *J. Hypertension*, 3 (Suppl. 3), S417, 1985.
51. **Kelbaek, H., Munck, O., Christensen, N. J., et al.,** Autonomic nervous control of postprandial hemodynamic changes at rest and upright exercise, *J. Appl. Physiol.*, 63, 1862, 1987.
52. **Onrot, J., Goldberg, M. R., Biaggioni, I., et al.,** Hemodynamic and humoral effects of caffeine in autonomic failure, *N. Engl. J. Med.*, 313, 549, 1985.
53. **Shannon, R. P., Wei, J. Y., Rosa, R. M., et al.,** The effects of age and sodium depletion on carotid vascular response to orthostasis, *Hypertension*, 8, 438, 1986.
54. **Ewing, D. J., Hume, L., and Campbell, I. W., et al.,** Autonomic mechanisms in the initial heart rate response to standing, *J. Appl. Physiol.,* 49, 809, 1980.
55. **Zoller, R. P., Mark, A. L., Abboud, F. M., et al.,** Role of low pressure baroreceptors in reflex vasoconstrictor responses in man, *J. Clin. Invest.,* 51, 2967, 1972.
56. **Wieling, W.,** Standing, orthostatic stress and autonomic function, in *Autonomic Failure*, Bannister, R., Ed., University Press, Oxford, 1988.
57. **Shepherd, J. T. and Mancia, G.,** Reflex control of the human cardiovascular system, *Rev. Physiol. Biochem. Pharmacol.*, 105, 1, 1986.
58. **Folkow, B., Grimby, G., and Thulesius, O.,** Adaptive structural changes of vascular wall in hypertension and their relation to the control of peripheral resistance, *Acta Physiol. Scand.*, 44, 255, 1958.
59. **Rowe, J. W. and Kahn, R. L.,** Human aging: usual and successful, *Science*, 237, 143, 1987.
60. **Wolthuis, R. A., Hoffler, G. W., and Johnson, R. L.,** LBNP as an assay technique for orthostatic tolerance, *Aerospace Med.*, 41, 29, 1970.

Chapter 5

SEX DIFFERENCES IN RESPONSE TO ORTHOSTATIC AND OTHER STRESSES

Mary Anne Bassett Frey and Carol M. Porth

TABLE OF CONTENTS

I. INTRODUCTION

As women ourselves, we wish this book didn't have a separate chapter on the responses of women; we wish it didn't need to have a separate chapter. We wish scientists and medical personnel knew as much about the responses of women as they do about those of men, so that the other chapters could discuss their topics with regard to all people. But published data about women's responses, to orthostatic or other stresses, are limited. Occasionally data from a few women are integrated in a study with the data from (usually many more) men, but then this deprives us of learning about women's responses and the similarities and differences between the sexes. Once a lack of difference between the sexes is established in a specific area of research, we can legitimately combine the data from men and women in that area and discuss "human responses".

Meanwhile, we appreciate this opportunity to review research on women's responses to orthostatic stress. Where we can, we will compare the responses reported for women to those reported for men. In some areas, data suggest some differences in magnitudes or primary mechanisms of responses. These offer intriguing areas for future research. We hope that this chapter will encourage you to learn more about the physiology of women — in the library, in the laboratory, or in the field.

In this chapter, since women have less morbidity and mortality from cardiovascular disease than men,[1] we will start with a short summary of basic cardiovascular physiology and cardiovascular diseases of women, which will also provide a base for the rest of the chapter. Next, we will discuss the cardiovascular responses of women and comparable responses of men to exercise and nonorthostatic stresses. We will then discuss women's responses to four orthostatic, orthostaticlike, or "hyperorthostatic" stresses: (1) sitting and standing, (2) head-up tilt, (3) lower-body negative pressure, and (4) centrifugation in the $+G_z$ vector. For each stress, we will describe data from studies reporting women's responses and comparisons with similar data from men; effects of menstrual cycle phases; effects of physical (especially aerobic) fitness on women's responses; changes in women's responses with aging; and effects of a preceding bedrest, head-down tilt, or water immersion on women's responses to the stress. In each section, where possible, we will discuss the data for women in terms of comparable data for men.

II. CARDIOVASCULAR PHYSIOLOGY OF WOMEN

The cardiovascular responses of both women and men to orthostasis and other stresses reflect the function of the arterial and low-pressure baroreceptor reflexes and the central nervous system; the responsiveness to hormones; the ability of the arterioles to control peripheral vascular resistance, the venules and veins to control compliance, and the heart to adjust its rate and force of contraction; the filling of the vascular compartment; and the adequacy of the intrathoracic blood volume. However, women, with their different body size and composition and different hormonal profiles, may recruit and utilize these mechanisms differently from men.

The typical man in his 20s is 4 in. taller and 29 lb heavier, and he has a heavier skeleton (29 vs. 15 lb) and less fat content (23.1 vs. 33.8 lb) than the typical woman of a similar age. These differences exist even when the amounts of fat, muscle, and bone are expressed as percentages of body weight.[2] Whether the difference in body fat is due to biological or behavioral causes, owing perhaps to the more sedentary lifestyle of the average women, is not known.

Furthermore, the maximum aerobic capacity of the average sedentary woman, even when corrected for body size, is 15 to 30% less than that of the average sedentary man.[3,4] This may be the result of a lesser total oxygen-carrying capacity of their blood resulting from lower blood volume, fewer red blood cells, or lower hemoglobin content; lesser lean body tissue; a more sedentary lifestyle; or to other reasons that will be discussed later in the chapter.

A. CARDIAC FUNCTION

Several differences exist between cardiac function in women and that of men. The resting heart rate of women is higher than that of men at all age levels.[5] Women have smaller hearts, with smaller stroke volumes. Therefore, their heart rates are higher for any given cardiac output or oxygen uptake. However, exercise training programs produce similar percentage increases in aerobic capacity for both sexes. Using electrocardiography, phonocardiography, and analysis of the carotid pulse contour on 221 healthy women and 361 healthy men, 10 to 69 years, Montoye et al.[6] observed that (1) heart rates of supine women were faster than those of supine men, (2) preejection periods of men and women were equal and increased with aging in both sexes, and (3) left ventricular ejection times, which were longer in women than men, decreased with age in men but increased with age in women. These data are consistent with data that suggest that men have greater decreases in stroke volume with age than women do.[7]

Studies using transthoracic electrical impedance (Z_0) measurements, which reflect the resistivity of blood, tissues, and air and the length and cross-sectional area of the thorax, indicate that women have a greater thoracic impedance than men. Frey et al.[8] reported that the Z_0 (mean ± SEM) of the women was greater than that of men based on a study of 35 women (mean age = 24.2 years, mean Z_0 = 31.5 ± 0.55 Ω) and 29 men (mean age = 25.6 years, mean Z_0 = 23.5 ± 0.30 Ω). These investigators attributed this greater Z_0 in women to their greater resistivity resulting from a smaller thoracic cross-sectional area, greater chest wall and breast tissue fat, smaller heart, and lesser central blood volume. McKinney et al.[9] observed similar differences in Z_0 between women (n = 19) and men (n = 19), ages 24 to 63 years (mean age 41.3 years), in both the seated (28.2 ± 0.7 Ω and 21.8 ± 0.6 Ω, respectively) and supine (26.7 ± 0.7 Ω and 21.2 ± 0.7 Ω, respectively) postures. In this study, the anthropometric variables of height, weight, percentage of body fat, and subscapular fat correlated highly with Z_0 when subjects were both seated and supine, but did not totally explain the observed sex differences. However, a second group of ten well-trained women (mean age = 25.6 years) studied by these same investigators had lower Z_0 when seated than did the previously studied female subjects (24.8 vs. 28.2 Ω). In another study, when Porth et al.[10] compared Z_0 values for younger (mean age = 21.8 years) and older (mean age = 43.0 years) groups of sedentary women (24.6 ± 0.7 Ω and 29.0 ± 0.7 Ω, respectively) with Z_0 values for younger (mean age = 24.6 years) and older (mean age = 43.3 years) groups of highly fit women (23.6 ± 0.6 Ω and 24.3 ± 0.6 Ω, respectively), they found that Z_0 was affected by aerobic fitness as well as by sex. Unfortunately, there were no measurements of fat or heart size for these groups of women. These values may also vary between studies because of electrode placement and other methodological differences. Therefore, absolute values for Z_0 are not being compared among the studies.

B. PERIPHERAL VASCULAR FUNCTION

Evidence also reveals differences between women and men in the blood vessels, vascular responsiveness, and sensitivity or density of vascular receptors. Using venous occlusion plethysmography as an indicator of vascular responsiveness, Freedman et al.[11] monitored finger blood flow after the infusion of three different types of adrenergic drugs via a catheter in the brachial artery, cold applied to the neck, and reactive hyperemia. Administration of adrenergic agonists elicited vascular responses in the men, but none in the women. Both phenylephrine (a selective α_1 agonist) and clonidine (a selective α_2 agonist) decreased finger blood flow in men proportionate to the dose, and isoproterenol (a nonselective β agonist) increased blood flow proportionate to the dose. Other stimuli (e.g., nitroglycerin, reactive hyperemia, or digoxin, all of which act directly on the blood vessels, or cooling of the neck or administration of tyramine, both of which act through neural secretion of norepinephrine) elicited similar responses in women and men. There were no sex differences in the heart rate or blood pressure (systolic or diastolic) change that could be used to explain these findings. Hence, the investigators attributed the differences in response to adrenergic agonists to women having lower sensitivity or lower density of peripheral vascular adrenergic receptors than men.

C. BLOOD PRESSURE

Most references report that women have lower systolic and diastolic blood pressures than do men of a comparable age. These differences could result from inherent sex differences in the heart, the blood vessels, the nervous or endocrine systems, or to psychological or sociological differences.

In the study of physiological states and responses of women, the question arises as to whether variables change with variations in hormone levels during the menstrual cycle. The menstrual cycle is usually divided into two phases: the follicular phase, which begins with the first day of menstrual flow and ends with ovulation (around day 14 for a 28-d cycle), and the luteal phase, which extends from ovulation until the next menstrual flow. Estrogen levels begin to rise sharply around days 8 to 10, and they peak just prior to ovulation; whereas progesterone increases from about day 17 to day 26. Several investigators have monitored women's blood pressures during a series of menstrual cycles. Kelleher et al.[12] monitored the blood pressures of 18 women (19 to 38 years) at the same time of day, three times a week, over an 8-week period of time. These investigators had no measures (e.g., hormones) of where in her cycle the subject was when each measurement was made, except for counting from the day that menses started. Systolic pressure rose gradually during the week prior to the first day of the menses, was at its highest on the day of onset, fell by day 3 to levels comparable with days 14 to 21, and did not rise at midcycle. The changes in diastolic pressure in this study were too small to be reported.

In a retrospective cross-sectional study of data from 207 nonpregnant white women (18 to 44 years), Greenberg et al.[13] found that the mean level of systolic pressure (120.2 mmHg) for the follicular phase of the cycle did not differ from the mean systolic pressure (122.5 mmHg) for the luteal phase, but it was higher (135.4 mmHg) for the part of the luteal phase (days 17 to 26) during which progesterone levels peak. Diastolic pressures did not vary during the cycle. An 8-week prospective study done by the same investigators failed to confirm the systolic pressure difference observed between days 1 to 14 (106 mmHg) and days 17 to 26 (106 mmHg) that was found in the retrospective study. However, diastolic pressures did differ ($p < 0.01$) between days 1 to 14 (65.1 mmHg) and days 17 to 26 (63.9 mmHg). Other studies of cyclic changes in women's blood pressure suggest that pressures increase at midcycle.[14,15]

Another study revealed some differences that affect women's and men's blood pressures on a minute-by-minute basis. James et al.[16] studied 67 female (mean age = 52 years) and 137 male (mean age = 48 years) patients with hypertension. Blood pressures of these subjects were monitored four times each hour throughout a day as the subjects performed their usual daily activities. Each subject also kept a diary, noting his or her posture, emotion, and situation at the time of each blood pressure measurement. Systolic pressure of the men was significantly affected by emotion and situation, with anger eliciting the highest pressures, followed by anxiety; their diastolic pressure was also affected by emotional state, with anger eliciting the highest pressures. Posture (contradicting our reports later in this chapter) did not affect the men's systolic or diastolic pressures. Systolic pressure of the women, on the other hand, was affected most by posture and secondly by situation. Their diastolic pressure was also primarily affected by posture but was secondarily affected by emotion. For the women, anxiety elicited higher pressures than anger did. The authors alerted us to the fact that the environments of their women and men subjects differed, i.e., most of the men were at work as professionals and managers, whereas most of the women were at home.

In a study that revealed possible physiological or pathological causes for female-male differences in cardiovascular function, Hughes et al.[17] measured ankle and brachial systolic blood pressures using a Doppler technique and measured the diameter of the terminal abdominal aorta using echocardiography in 97 "normal" men and 101 "normal" women (20 to 90 years).

They observed that systolic arterial pressure at the ankle decreased with age in the men, but not in the women, and that systolic arterial pressure in the brachial artery increased more with age in women than in men. Diastolic pressure in the brachial artery did not change with age in either sex. Thus, the ankle-to-brachial pressure ratio decreased with age, due to different mechanisms in these women and men. The investigators attributed these differences to sex-mediated modifications of the arterial walls in the upper and lower limbs and suggested that sex hormones may affect pulse-wave transmission.

The "cause" of hypertension continues to be an enigma to medical researchers, enticing investigators to identify correlates of changes in blood pressure that might provide clues to its etiology. Daniels et al.[18] identified correlates of changes in blood pressure of a large population of black and white men and women in Evans County, GA, who were studied from 1960 to 1967. Of the 3102 initial participants in the study, 2530 were retested. These investigators compared their data among the four sex/race groups of men and women. Thus, while a racial comparison is not an objective in this chapter, we are, in this instance, including the data as reported. Systolic blood pressure in all four sex/race groups and diastolic pressure in all groups except black women were positively correlated with a change in the Quetelet index of ponderosity ($100 \times$ weight [lb]/height [in.]). Systolic pressure in the white men only and diastolic pressure in all four groups were negatively correlated with age. The interesting aspect of the study for this chapter is the large number of positive correlations that were reported as significant only for the group of white women. These included correlations of change in systolic pressure with baseline systolic pressure, baseline Quetelet index, level and change in socioeconomic status, and change in cholesterol; as well as correlations of change in diastolic pressure with the baseline Quetelet index, and the level and change in socioeconomic status.

III. CARDIOVASCULAR DISEASE IN WOMEN

Prevalences of cardiovascular disease processes, particularly coronary artery disease (CAD), differ between men and women. Even the interpretation of diagnostic tests of men and women may vary. Because heart disease is more prevalent in men, most interpretations of cardiac diagnostic testing and risk factors have been determined largely from studies on men.

Observations based on the Framingham Heart Study indicate that CAD accounts for 23% of all deaths in women and 34% of deaths in men.[19] Furthermore, clinical evidence of CAD appears, on average, 10 years later in women than it does in men, and myocardial infarction occurs 20 years later in women than it does in men.[20]

Angina pectoris is reported to be the initial manifestation of disease in 55% of women who develop CAD as compared with 43% of men;[21] however, some studies have shown that among women with typical angina, only 50% have significant CAD.[20] A study by Boucek et al.[22] reported that, in a study of 100 women and 100 men (65 years or older) with angina, coronary arteriography indicated that men had a higher incidence of stenosis (greater than 70%) of the left main and left circumferential arteries and poorer left ventricular perfusion than women. However, 20% of both the men and the women in the study had one-vessel obstructive disease. Furthermore, the Coronary Artery Surgery Study (CASS) data revealed a 2.7 times greater operative mortality with coronary artery bypass grafting for women than for men, despite the presence of less severe coronary disease and of better ventricular function.[23] In attempting to understand the sex differences in the pathology of CAD, the CASS investigators examined the influence of the size of coronary arteries on the success of the surgical outcome. Their results showed that the women had smaller coronaries and that this was closely related to mortality.[23]

First heart attacks are more often fatal in women (39%) than in men (31%).[20] Furthermore, the risk factors for sudden unexpected death (death occurring within 1 h of the onset of symptoms in persons without prior overt CAD or without other probable cause) are reportedly different for women than men. In the Framingham Study the risk factors for CAD in men (smoking, hypertension, excess weight, and elevated blood cholesterol levels) were the same as those for sudden unexpected death (SUD), suggesting a strong relationship between the atherosclerotic process and SUD.[24] In the women, these CAD risk factors were not correlates of SUD. Instead, vital capacity (decreased) and hematocrit (increased) were predictors of SUD in this group of women.[24] The hematocrit finding is of interest for two reasons: one is the lower hematocrit during the premenopausal period, when the incidence of CAD is less in women as compared with men, and the second is the increasing attention being given to hematological factors in CAD.[25,26] Shatzkin et al. suggested that hematocrit may act directly via viscosity effects, or it may be a proxy for some other hematological event such as an increased coagulable state.[24]

The relatively low incidence of CAD in women may also render invalid the diagnoses using noninvasive methods that have been developed based on criteria from populations of men with and without disease. The sensitivity of exercise testing is influenced by the extent and severity of CAD. Since women are less likely to have severe CAD, a positive exercise test is more likely to be a false positive in women than in men. Detry et al.[27] reported false-positive results in 38% of women and false-negative results in 40% of men in a sample of 47 women and 231 men with complaints of angina (typical or atypical) or a history of myocardial infarction. They found that when both the history (typical angina) and the exercise ECG were positive, 98% of the male patients had CAD as diagnosed by angiograms, and when the history (atypical angina) and exercise ECGs were both negative, 89% of the male patients were negative for CAD by angiograms. However, 25% of women with positive exercise tests and histories were free of CAD as diagnosed by arteriography. In a study by Sketch et al.,[28] in which 195 men and 56 women with chest pain underwent multistage submaximal exercise testing, positive exercise tests were associated with no stenosis or less than 50% stenosis in only 8% of the men as compared with 67% of the women. Conversely, the incidence of false-negative exercise tests associated with 75% stenosis or greater was higher in men (37%) than women (12%).

The explanation for this high prevalence of false-positive exercise tests in women is unclear. Kusuml and Bruce have suggested that postexertional S-T segment depression may be partly due to lower levels of hemoglobin and total circulating red cell mass.[29] Furthermore, Kusuml and Bruce indicate that asymptomatic women who exhibited S-T depression after maximal exercise had greater pulmonary and systemic resistances with higher pressure-rate products. Other factors that could have influenced test results were the reported higher incidence of mitral valve prolapse and S-T and T wave electrocardiograph changes and the finding that 30% of women did not increase their ejection fraction during exercise testing as the men did.[30] Estrogen could be partially responsible for the high incidence of false-positive test results in women. Estrogen and digitalis have similar chemical structures, and digitalis glycosides are known to be associated with an increased incidence of false-positive test results.[28]

IV. WOMEN'S RESPONSES TO EXERCISE AND NONORTHOSTATIC STRESSES

Documented responses of women to exercise and other nonorthostatic stresses and comparison of these responses with those of men may provide a fertile basis for understanding similarities and differences between the sexes in response to orthostatic stresses. We have therefore reviewed reports of women's responses to dynamic exercise and several nonorthostatic stresses, including cold, isometric and dynamic handgrip, the Valsalva maneuver, and mental tasks.

A. EXERCISE
We are not endeavoring to report fully on the responses of women and men to exercise stress. We are, however, including data from several studies that have compared the responses to exercise of men and women over a wide age range. Through these, we hope to present a full

picture of the cardiovascular physiology of "normal" women as compared with that of "normal" men. Two sex differences consistently observed in these studies are (1) greater stroke volume, cardiac output, and oxygen uptake of men, and (2) increases in stroke volume and ejection fraction of men from submaximal activity to maximal exercise, but no such changes in women.

Women have, on the average, lower maximal oxygen uptake (Vo_{2max}) and physical work capacity than men. Maximal oxygen uptake (reported in ml/kg/min) reflects largely the ability of the cardiovascular system to deliver oxygen to the working muscle mass. Because women have relatively more fat, the differences between the sexes are reduced when oxygen uptake is expressed as relative to fat-free body mass. In a large study of 375 women and 1514 men (17 to 55 years), composed of new recruits entering the army from civilian life and soldiers on a variety of assignments and physical training programs, the Vo_{2max} during discontinuous treadmill exercise (using a modified Taylor protocol) was 30% less for women than men when expressed on an absolute basis and 14% less than that of men when expressed as a function of fat-free weight.[31] In this study, the men's aerobic capacity decreased with age at an average yearly rate of 0.5 ml/kg/min. In a longitudinal study by Astrand et al.,[32] the rate of decline of Vo_{2max} was recorded as 0.44 ml/kg/min/year in 35 women who had been active for most of 35 years, and this rate was less than that reported for a similar age group of men. Other studies have shown that the Vo_{2max} for men is approximately 50% greater than that of women when expressed in l/min, 15 to 25% greater relative to body weight, and 5 to 15% greater relative to fat-free weight.[33,34]

Astrand et al.[35] monitored the responses of 11 female and 12 male well-trained physical-education students (ages 20 to 31 years) during maximal and 50% submaximal (ergometer) exercise (see data listed in Table 1). The women had smaller stroke volumes at rest (68 vs. 88 ml) and during exercise than did the men, and, unlike the men, their stroke volume during maximal exercise was no greater than that during submaximal exercise. The authors reported high correlations of maximum stroke volume, cardiac output, and oxygen uptake with plasma volume and heart rate. The greater cardiac output per liter of oxygen used for the women was reportedly related to their lower hemoglobin concentration.

Hossack and Bruce,[36] who studied 104 normal sedentary women (20 to 70 years) and 98 normal sedentary men (20 to 73 years) during treadmill tests (Bruce protocol), found that the men, in comparison with the women, had larger maximal values for cardiac index and stroke index (estimated by regression from the measured value of oxygen consumption in individual subjects), as well as greater oxygen content in the blood and greater oxygen extraction from the blood. The estimated stroke index of their female subjects did not change from rest to maximal exercise, whereas that of the men increased. These women also had higher blood pressures, both at rest and at maximal exercise, and greater peripheral resistance at maximal exercise than did the men.

Sixteen healthy women (32 to 68 years) and 15 healthy men (31 to 67 years) performed upright ergometer exercise as they were studied by Higginbotham et al.,[37] who used quantitative radionuclide angiography and an analysis of expired gas. The women's and men's ejection fractions were equal when they were at rest, but when they were at maximal exercise, the ejection fraction of the men increased from a resting value of 0.62 to 0.77, whereas that of the women remained at 0.64. During exercise, stroke volume was increased in the women because of an increased end-diastolic volume, whereas the men had a reduced end-systolic volume, possibly indicating that their contractile reserve was greater. This investigation differed from others in that the sexes had similar maximal oxygen uptake, oxygen extraction, and systolic blood pressure. However, the investigators did note that 3 of their 15 male subjects were obese.

Adams et al.[38] used radionuclide ventriculography to study 28 asymptomatic women (mean age = 33 years) and 27 asymptomatic men (mean age = 30 years) as the subjects performed ergometer exercise in the supine posture until they were fatigued. The women and the men had similar heart rates, but the men had higher blood pressures. These investigators also observed

TABLE 1

**Mean Values for Cardiovascular Variables and Oxygen Uptake
of Men and Women During Maximum and Submaximum Exercise**

Variable	Men (n = 12)		Women (n = 11)	
	Max	Submax	Max	Submax
Heart rate, bpm	186	124	194	137
Stroke volume, ml	134	125	100	93
Cardiac output, l/min (dye dilution)	24.1	15.4	18.5	12.6
Oxygen uptake, l	4.05	2.03	2.60	1.30
CO/l O_2 used, l/min/lO_2	5.9	7.6	7.1	9.7
A-V O_2 difference, ml/l	17	13.2	14.3	10.3

Data from Astrand, P. O., Cuddy , T. E., Saltin, B., and Stenberg, J., *J. Appl. Physiol.,* 19, 268, 1964.

that, although the women and men had equal ejection fractions at rest (0.64), the men's ejection fraction increased significantly more than the women's did during exercise, because of the men's reduced end-systolic volume. These results differed from those of Higginbotham et al.[37] in that the end-diastolic volume of these male subjects decreased during exercise, and the end-diastolic volume of these female subjects did not change. This difference between studies could have resulted from the postures of the subjects: supine in this study and upright in the Higginbotham et al. study.

Women on the average have a blood hemoglobin concentration that is approximately 15% lower than that of men.[38] This lower hemoglobin concentration results in a lower oxygen content of the blood and may contribute to the lower Vo_{2max} of women. Cureton et al.[39] studied 10 men and 11 women as they cycled on an ergometer using a continuous, load-incremented protocol. The men were tested first with their normal hemoglobin levels and were retested after the blood was withdrawn to reduce their hemoglobin to be exactly equal to that of the group of 11 women. Prior to blood withdrawal, the Vo_{2max} (ml/kg/min) of the men was 11.5% greater than that of the women, and their exercise time was 67% longer. However, neither maximal heart rate nor Vo_{2max} (adjusted for fat-free weight) differed between the sexes. Equalizing the hemoglobin reduced the Vo_2 by only 6.9% and ride time by only 4.8%, less than would have been predicted by proportional changes in the arterial blood and arteriovenous oxygen content difference. These findings led the investigators to conclude that hemoglobin accounts for a significant, but relatively small, portion of the sex differences during maximal exercise and that other factors, such as the dimensions of the oxygen transport system and musculature, are of greater importance.

Unfortunately, few of the studies that measured maximum oxygen uptake reported activity levels of the subjects. The average "normal" woman and even the average "sedentary" woman may exercise less on a routine basis in an average day than the "normal" or "sedentary" man. Until it is determined whether this assumption is true, we will not know whether the greater maximum oxygen uptake usually reported for men is entirely a physiological sex difference or whether it results in part from differences in activity levels.

The changes with age in several of these cardiovascular variables also differ quantitatively between men and women, and the results of the several studies reported are consistent in this regard: maximum heart rate decreased with age, but more for the men than for the women;[36] maximum estimated cardiac index also decreased with age, more for the men than for the women;[36] ejection fraction likewise decreased with age, but again, more for the men than for the women.[37]

Convertino and colleagues[40,41] compared the responses of men and women in young age groups (men 19 to 23 years and women 23 to 34 years) and older age groups (mean age = 55 years) to supine exercise on an ergometer before and after 10 to 17 d of bedrest. The men had greater maximum oxygen uptake. Plasma volume and maximum oxygen uptake for all age groups decreased after bedrest, and maximum heart rate increased in all groups. The only difference between the sexes was a decrease after bedrest in the oxygen uptake at submaximal exercise in the group of younger women, but not in the group of younger men. The authors indicated that the mean weight of these women was lower after bedrest because of an intentional loss by two of the subjects; this could have reduced the women's oxygen uptake.

B. COLD

Exposure of the body to cold presents a physiological stress. Responses to cold have been measured during exposure of the whole body or exposure of portions of the body, such as the limbs or face. In one study of exposure of the whole body, four groups of subjects (10 women and 10 men [20 to 30 years] and 7 women and 10 men [51 to 72 years]) spent 2 h each in temperature chambers at 28, 20, 15, and 10°C. They were positioned in a semisupine posture and dressed in only bathing suits.[42] Arterial pressures of the women and those of the men did not differ. Diastolic pressure increased in the cold, while systolic pressure did not change. Cardiac output was not changed in any of the conditions, except the last hour at 10°C for the men only. Neither the women's heart rate nor their stroke volume changed with exposure to cold, but the men's heart rate decreased and their stroke volume increased. The men may have had greater cutaneous vasoconstriction, shifting more blood to their central circulation; they may have had greater sympathetic outflow to the heart itself, increasing stroke volume due to greater contractility; or they may have had both.

When six women and six men (22 to 41 years) performed cold-pressor tests with water at 4°C in four conditions (hand or foot immersion when seated or supine), both systolic and diastolic pressures increased throughout all immersions.[43] The women's arterial pressures did not differ from the men's before the immersions; but systolic pressure of the women increased less than that of the men did during two of the tests (seated hand immersion and supine foot immersion), and the diastolic pressure of the women increased less during supine foot immersion. These results agree with those from a study in which eight women (mean age = 30.4 years) and nine men (mean age = 35.0 years) immersed their hands in water at 5°C for 2 min.[44] Arterial pressures of the women and men did not differ before hand immersion, but, in response to exposure to the cold water, those of the women increased less than those of the men did. During the first minute, the women's heart rates increased more than the men's did. However, in another cold-pressor study, responses in systolic pressure, diastolic pressure, and heart rate to 90 s of foot immersion in ice water did not differ between a group of 24 women and a group of 12 men;[45] nor did responses in blood pressure or heart rate differ between women in the follicular phase (days 7 to 11) and women in the luteal phase (days 17 to 21) of their menstrual cycles.

During facial exposure to 66-km/h wind at 0°C, heart rates decreased similarly for eight women and nine men.[44] Heart rates also decreased in another study in which the faces of ten women (mean age = 24.1 years) and ten men (mean age = 24.7 years) were exposed to 5.0 m/s wind at 4°C and 50% relative humidity.[45] These investigators also observed that systolic and diastolic pressures increased and forearm blood flow decreased during exposure of the face to cold, with no differences between the sexes.

In summary, these studies suggest that cold to any part of the body of either women or men elicits an increase in arterial pressure. Otherwise, responses and differences in responses between the sexes depend on which part of the body is exposed. With facial exposure, heart rate decreases, apparently with little or no difference between the sexes. Limb exposure elicits less arterial pressure increase in women than it does in men. The study of Wagner et al.[42] suggests some sexual differences in how cardiac output is maintained during exposure to cold.

C. ISOMETRIC AND DYNAMIC HANDGRIP EXERCISE

Although handgrip exercise involves only a small portion of the body's musculature, it elicits cardiovascular responses greater than required for the increased metabolism. Petrofsky et al.[47] examined strength, endurance, and cardiovascular responses to isometric handgrip in 83 women (19 to 65 years) and compared these data with those from 100 men they had studied previously. Strength of the women was 28.9 ± 6.1 kg, which was only 58% of the men's strength, while their endurance time at 40% maximum voluntary contraction was 172 ± 52 s, which was greater than that of the men (139 ± 31 s). Arterial pressures of the women were lower than, and increases in arterial pressure during isometric handgrip were less than, those of the men. Heart rate increased equally in both sexes. The women did not lose strength and endurance with aging, but the men did.

Kobryn et al.[48] compared the responses of 20 women and 20 men (20 to 39 years) to dynamic handgrip, consisting of pumping a dynamometer 60 times per minute at 40, 60, and 80% maximum voluntary contraction. Systolic pressure of the women increased less than systolic pressure of the men did, as with isometric handgrip. In this test, however, the heart rate of the women increased less than the heart rate of the men did, and the women appeared to have shorter endurance times to exhaustion.

As with their responses to cold exposure, women appear to have less arterial blood pressure elevation with handgrip stress than men have.

D. VALSALVA MANEUVER

The Valsalva maneuver produces transient changes in intrathoracic and intraabdominal pressures that challenge the cardiovascular system with rapid and transient changes in preload and afterload stresses. Venous return to the heart is decreased during the strain, and it is increased following the release of the strain. During the strain, stroke volume output and pulse pressure fall, which stimulates a baroreflex-mediated increase in heart rate and peripheral vascular resistance. Following release of the strain, a reflex bradycardia indicates a transient overshoot of arterial pressure above normal values. Booth et al.[49] measured the responses in heart rate, brachial artery blood pressures, right atrial pressures, cardiac output (bromsulphalein indicator dilution), and total peripheral resistance during a 40-mmHg Valsalva maneuver held for 40 s. They compared the responses of 11 women with those of 10 men. The heart rate and blood pressure changes of the women and men in this study did not significantly differ. However, the women had a greater decrease (-17.3 ± 6.2% of control) in central blood volume at 25 s of strain than the men had, and they had a significantly greater increase in cardiac index (+23.8% of control) following the release of the strain than the men had (–0.05% of control). These data suggest that women may sequester more blood in their extrathoracic veins during periods of increased intrathoracic pressure such as that associated with the Valsalva maneuver.

E. MENTAL TASKS

Mental tasks can also provide stresses to the cardiovascular system. Thus, perhaps psychological factors, and not exclusively sex-related physiological factors, determine some responses to stress. In a study of 12 women and 12 men (17 to 30 years), Frey et al.[50] observed that the women had smaller elevations in arterial pressure during mental arithmetic tasks, but heart rates of the women increased more than did heart rates of the men.

Hastrup et al.[45] used a shock-avoidance reaction-time task to challenge 24 women (12 in the follicular phase and 12 in the luteal phase of their menstrual cycles) and 12 men. The women who were in the luteal phase of their menstrual cycles and the men experienced greater increases in systolic pressure during stress than did women in the follicular phase. The men had greater increases in heart rate during stress than did the follicular-phase women, and the luteal-phase women tended to have greater increases in heart rate than did the follicular-phase women. Reaction times did not differ between groups.

During the Stroop color test, heart rates of six women (21 to 26 years) and six men (23 to 30 years) increased equivalently; but systolic pressure of the women increased less than did that of the men,[51] possibly because of the lower epinephrine levels in the women. The norepinephrine levels of the women were also lower than those of the men, but this difference was not significant at $p < 0.05$.

Differences in catecholamine secretion may explain in part the lesser blood pressure response of women to stress. Frankenhouser reported that, while the general pattern of catecholamine secretion is similar for women and men when they are resting, women secrete less catecholamines during stress.[52] A group of women (18 to 35 years) whom she studied did not secrete epinephrine during intelligence tests, color-word tests, or venipuncture, whereas a group of men did. During more stressful activities, the women did secrete catecholamines, but less than the men secreted. The women performed as effectively as or better than the men in all of these situations. She stated that when women assume roles that traditionally have been men's, e.g., that of an engineering student, their epinephrine secretion during stress becomes more like that of men.

In summary, women generally respond to stress, both physical and mental, with lesser elevations in blood pressure. This may be related to sex differences in catecholamine secretion. Some data indicate that during the luteal phase of the menstrual cycle women's blood-pressure responses to mental stress are similar to those of men and greater than their blood-pressure responses during the follicular phase of the menstrual cycle. With aging, women lose less maximal oxygen consumption, heart rate and cardiac index, ejection fraction, and isometric strength and endurance than men lose. The responses to exercise and nonexercise stresses are summarized in Table 2.

V. WOMEN'S RESPONSES TO "ORTHOSTATIC" STRESSES

In this section, data on the responses of women to four of the most studied types of orthostatic or orthostatic-like stresses are summarized: sitting and standing, head-up tilt, lower body negative pressure, and centrifugation to produce G stresses in the positive Z axis ($+G_z$). These data will be compared with those from men and with the other information about cardiovascular function of women discussed previously.

A. SITTING AND STANDING
1. Circulatory Responses
Over 60 years ago, Turner studied women's responses to sitting and standing.[53,54] In the first of these studies, Turner described metabolic, cardiac output (measured by the Fick carbon-dioxide method), and heart rate responses to changes in posture.[53] Twenty-five women (17 to 37 years) performed several experiments in which they were either (1) supine for 30 to 45 min then sitting for 30 min or (2) supine for 30 to 45 min, sitting for 15 min, and then standing for 8 to 14 min. The metabolic rate of the women increased 6% from supine to sitting and increased 13.6% from supine to standing. Despite the increase in metabolic rate, however, cardiac output decreased from 6.26 l/min when subjects were supine, to 5.59 l/min when they were sitting (10% decrease), to 4.77 l/min when they were standing (21.2% decrease). Stroke volume decreased from 96.9 ml when subjects were supine to 79.0 ml when they were sitting and to 56.6 ml when they were standing (a 40% decrease), and heart rate increased from 64.7 bpm when subjects were supine to 71.8 bpm when they were sitting and 85.1bpm when they were standing (33.4% increase). The cardiac output of some subjects declined progressively throughout the duration of standing, decreasing 39.6%; heart rates of these subjects rose conspicuously. Turner considered this decrease in cardiac output upon standing, which was concomitant with an increase in metabolic rate, to be a "less perfect adjustment" to standing.

In her second report, Turner characterized differences between subjects who had good

TABLE 2
Summary of Responses to Exercise and Nonorthostatic Stresses

Stress	Cardiovascular Responses	Sex Differences
Exercise	Increased heart rate,[35] increased stroke volume, increased oxygen uptake, increased AV O$_2$	Only men's stroke volumes increased from submaximal to maximal work.[35,36] Only men's ejection fractions increased or increased more.[35,36] Men had a larger maximal cardiac index and stroke index, and oxygen content.[36] Women had higher maximal arterial pressures and peripheral resistance. Only men's end-systolic volumes decreased.[37,38] Maximal heart rate, cardiac index,[36] and ejection fractions decreased more with age for men than women.[37]
Cold (whole body)	Increased diastolic pressure[42]	Stroke volume increased and heart rate decreased in men only.[42]
Cold (limb)	Increased systolic pressure[43,44]	Arterial pressures increased more in men;[43,44] heart rate increased more in women at minute 1.[44]
Cold (face)	Decreased heart rate,[44,46] increased diastolic pressure,[46] and decreased forearm blood flow[46]	
Handgrip (isometric)	Increased arterial pressure and heart rate[47]	Men had a greater increase in arterial pressure[47]
Handgrip (dynamic)	Increased arterial pressure and heart rate[48]	Men had a greater increase in systolic pressure and heart rate[48]
Valsalva	Decreased pulse pressure and increased heart rate during strain[49]	Women had a greater decrease in central blood volume at 25 s of strain[49]
Mental arithmetic	Increased arterial pressure and heart rate[50]	Men had a greater increase in arterial pressure and women had a greater increase in heart rate[50]
Reaction time	Increased systolic pressure[45]	Men had a greater increase in systolic pressure and heart rate than follicular-phase women, but were not different from luteal-phase women[45]
Stroop color test and IQ test	Increased systolic pressure, heart rate,[51] and catecholamine levels[52]	Men had a greater increase in systolic pressure[51] and epinephrine levels[51,52]

tolerance to orthostatic stress, which she defined as a decrease in cardiac output of less than 28% upon standing (mean decrease = 9.5%) and those who had poor orthostatic tolerance (mean decrease = 40%).[54] Twenty-four of the same women, 15 with good tolerance and 9 with poor tolerance, who had participated in the previously described study performed a protocol of 12 to 15 min reclining, 15 min standing, and another 10 min reclining. After 3 min in the standing position, the average heart rate of the group of subjects with "good orthostatic tolerance" increased 18 bpm, from a supine rate of 60.8 bpm, and the average heart rate of the group of subjects with "poor orthostatic tolerance" increased 19.2 bpm, from a supine heart rate of 69.8 bpm. The subjects with poor orthostatic tolerance had a progressive fall in systolic pressure during the period of standing, whereas the subjects with good orthostatic tolerance increased their diastolic pressure and maintained their systolic pressure when they stood. Turner con-

TABLE 3
Mean Values for Cardiovascular Variables of Women (n = 7)
in Supine, Seated, and Standing Postures

Variable	Supine	Seated	Standing
Heart rate, bpm	67	76	89
Stroke volume, ml	118	72	53
Thoracic impedance, Ω	31.0	34.4	36.2
Electromechanical systole, ms	410	372	342
Ejection time, ms	314	263	226
Pre-ejection period, ms	96	109	115
PEP/ET	0.307	0.420	0.517

Data from Frey, M. A. B., Doerr, B. M., Mann, B. L., and Miles, D. S., *Proc. IEEE NAECON,* 1086, 1981.

cluded that the rate of return of blood to the heart is the most essential feature in a successful circulatory response to an orthostatic challenge.

In a more recent study, Frey et al.[55] used electrocardiography, impedance cardiography, and phonocardiography to monitor the heart rate, thoracic impedance, electromechanical systole, preejection period, and ejection time of seven women (mean age = 31 years) during random assignment to the supine, seated, and standing positions. All measured variables, which are listed in Table 3, differed significantly among the postures. These results agree with and supplement Turner's data.

2. Comparison of Women's and Men's Responses

Moore et al. measured heart rate and blood pressure responses of 25 women and 25 men (25 to 35 years) during the first 2 min after they rose from supine to standing.[56] When supine, the women had a higher heart rate (67.4 vs. 60.5 bpm), lower systolic blood pressure (102 vs. 111.5 mmHg), and similar diastolic blood pressure (71.2 vs. 75.8 mmHg) than the men had. The changes on standing did not differ between men and women for any of these variables: heart rate increased approximately 12 bpm, systolic blood pressure decreased approximately 12 mmHg, and diastolic blood pressure increased 3 to 4 mmHg.

On the other hand, when Frey et al.[7] monitored 22 men and 25 women (21 to 59 years) in the supine, seated, and standing postures, using electrocardiography, impedance cardiography, and sphygmomanometry, they observed differences between women and men in all the measured variables and in the women's and men's responses to the change of posture. The men had higher stroke volume, cardiac output, systolic blood pressure, and diastolic blood pressure; and the women had higher heart rate, total peripheral resistance (supine only), and thoracic impedance. The women experienced a greater increase in heart rate between sitting and standing than between supine and sitting, as was also observed in the study described above.[55] In addition, several other differences were also observed between the sexes in their adjustment to postural change. Stroke volume (absolute value) decreased more between the supine and standing postures in the men than it did in the women, whereas peripheral vascular resistance and thoracic impedance increased more in the women than they did in the men. All variables, except systolic blood pressure, also differed among postures, with greater differences between the supine and sitting postures than between the sitting and standing postures for all variables and both sexes, except heart rate in the women. The values of the variables in the supine and standing postures are listed in Table 4, and several of these are graphed in Figure 1. These data illustrate a greater use of heart rate and peripheral resistance mechanisms to maintain blood pressure by women, who have smaller hearts and smaller stroke volumes than men have.

TABLE 4
Mean Values for Cardiovascular Variables in Women (n = 25)
and Men (n = 22) in Supine and Standing Postures

Variable	Supine		Standing	
	Men	Women	Men	Women
Heart rate, bpm	59	66	72	82
Stroke volume, ml	118	82	72	48
Cardiac output, l/min	6.8	5.4	5.2	4.0
Systolic pressure, mmHg	120	108	119	110
Diastolic pressure, mmHg	77	70	84	78
Mean arterial pressure, mmHg	91	83	96	88
Thoracic impedance, Ω	22.2	30.4	25.6	34.3

Note: All variables differed (p <0.05) between men and women. All variables differed (p <0.05), except systolic pressure, between postures.

Data from Frey, M. A. B., Tomaselli, C. B., and Freeman, M., submitted.

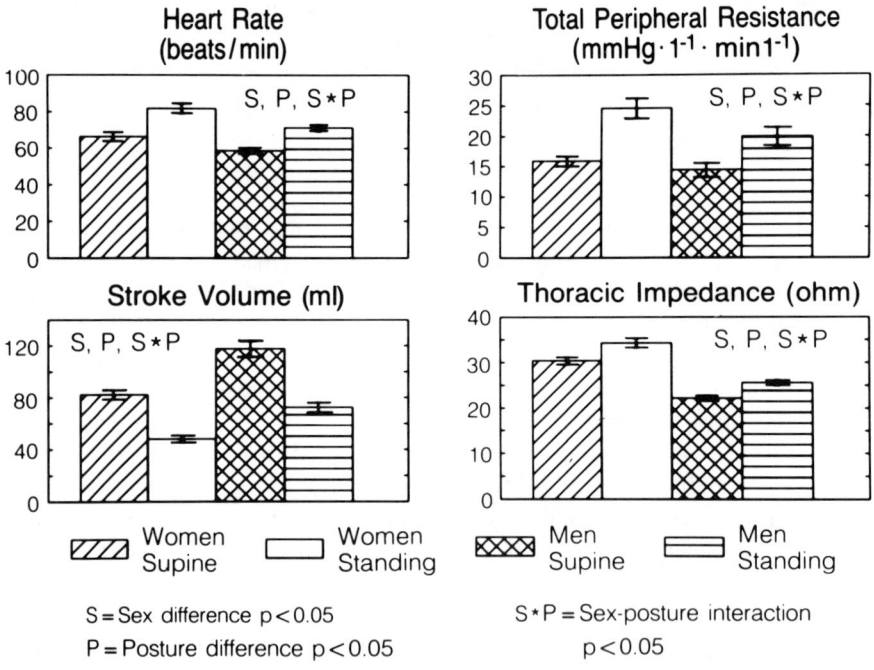

FIGURE 1. Values of cardiovascular variables in women and men.

3. Changes with Aging

Stroke volume and cardiac output decreased less in older subjects of both sexes as they rose from the seated to the standing position than they did with younger subjects, more so for men.[7] We previously described observations that stroke volume and ejection fraction decrease more with aging in men.[7,37] However, all these studies leave unanswered the question of whether stroke volumes of men decreases more with aging because the activity level of men decreases more with age or because of changes, pathological or otherwise, in the heart or blood vessels.

Total peripheral resistance and thoracic impedance were positively correlated with age in both sexes, with no differences between the sexes.

B. HEAD-UP TILT
1. Circulatory Responses

In 1930, Turner reported on several series of experiments intended to reveal why the blood of some subjects was not effectively returned to their hearts when they were standing.[57] In this study, three series of tests were used to compare responses to passive head-up tilt with responses to standing, which requires more activity by the subject. In one series, 12 women (16 to 30 years) performed repeated tests in which they went from a supine posture to either 62° passive head-up tilt (an angle that was chosen because it would elicit most of the orthostatic stress of standing without the activity of standing) or 90° head-up standing. With passive head-up tilt, oxygen uptake increased 5.8%, heart rate increased 30%, and pulse pressure decreased 20%; with standing, oxygen uptake increased 19%, heart rate increased 28%, and pulse pressure decreased 24%. These data indicate that the activity involved in standing causes the metabolic rate to increase, while orthostasis causes heart rate to increase. In the next series, 11 women were monitored at seven tilt angles from horizontal to 90° head up. Their heart rates increased progressively from 62 to 88 bpm as the angle of tilt was increased, while their pulse pressures decreased progressively from 39 to 26 mmHg. In the final series, measurement of leg volumes by water plethysmography revealed that only 4 to 5% of the subjects' blood was pooled in their legs when they stood. These tests also indicated that a tight bandage or a corset around the subjects' abdomens when they were standing improved their circulatory function. From these results, the investigator concluded that blood accumulates not only in the legs, but also in the abdomen when subjects stand.

2. Comparisons of Women's and Men's Responses

Shvartz and Meyerstein compared the responses of 18 women (17 to 23 years) with those of 18 men (17 to 32 years) when the subjects were exposed to 20 min of 70° head-up tilt after they had been horizontal for a 30-minute period.[58,59] As shown in Table 5, the absolute values of all variables, except pulse pressure, differed between the women and the men in each posture, but changes in the variables with change of posture did not differ between the sexes. One man and one woman fainted before the end of the 20-min tilt.

When subjects were tested in four conditions — hot room at 40°C and temperate room at 24°C, in winter and summer (in Israel) — more syncopal episodes and symptoms of intolerance to orthostatic stress occurred in the heat tests during winter than in the other conditions, but no significant differences existed between the sexes in the occurrence of fainting, symptoms of orthostatic intolerance, heat acclimation, or responses to tilt.

3. Responses After 12 Days of Acclimation to Heat

In another study, six women (mean age = 22 years) and four men (mean age = 27 years) were tilted to 70° head up before and after 12 days of acclimation to heat (by exercise on an ergometer in a chamber at 40°C, 42% relative humidity for 2 h/d).[60] After the acclimation period, plasma volume and blood volume of both the women and the men were increased, and hemoglobin and hematocrit were decreased. Blood pressures during tilt did not differ between the sexes or between the preacclimation and postacclimation tilt tests. However, the number of subjects in this study may have been too small to reveal a difference. The heart rates of the women during tilt did not change from before to after acclimation; however, the heart rates of the men during tilt were lower after acclimation than they had been before. Also, the time for which men could tolerate tilt increased after they had been acclimated to the heat, whereas the women's tolerance time to tilt did not change. The investigators have no firm explanation for this difference between the men and the women, but suggest the women's inability to improve orthostatic tolerance time after acclimation may be due to their smaller plasma volume.

TABLE 5
Mean Values for Cardiovascular Variables of Men (n=18)
and Women (n = 18) in the Supine and 70° Head-Up
Tilted Position

Variable	Supine		Tilted	
	Men	**Women**	**Men**	**Women**
Heart rate, bpm	68	75[a]	86	91[a]
Systolic pressure, mmHg	114	107[a]	111	105[a]
Diastolic pressure, mmHg	75	73[a]	87	82[a]
Pulse pressure, mmHg	39	34[a]	24	23

[a]   Men and women differ, $p < 0.01$

Data from Schvartz, E. and Meyerstein, N., *Aerospace Med.*, 41, 253, 1970.

4. Effects of Fitness and Changes with Aging

Porth et al.[10] studied the effects of age and fitness level on response during 5 min of graded (10, 30, and 70°) head-up tilt. Fitness level was based on Vo_{2max}: young sedentary less than 23 ml/kg/min, young fit greater than 49 ml/kg/min, older sedentary less than 21 ml/kg/min, and older fit greater than 41 ml/kg/min. The subjects were tested in the middle 2 weeks of their menstrual cycles. Table 6 describes the circulatory responses of the women in this study to 70° tilt. The supine heart rate for the women in the young and older fit groups was significantly less ($p < 0.01$), and the supine stroke volume index for these groups was significantly greater ($p < 0.01$) than for the sedentary groups. However, the percent change from control values for these variables with tilt did not differ among groups for any of the tilt levels. Cardiac index of the young fit women was significantly greater than that of the older sedentary women at 30°, and total peripheral resistance of the young fit women increased less at the higher levels of tilt (30° and 70°). These findings are consistent with other data described earlier in this chapter, indicating that cardiac index decreases and total peripheral resistance increases with aging in men and women, and with the generally accepted view that aerobically fit individuals have greater cardiac indices.

In summary, the limited data available reveal few differences between women and men in their responses to head-up tilt. Men, however, may improve their ability to withstand orthostatic stress with a heat/exercise training program more than women do.

C. LOWER-BODY NEGATIVE PRESSURE

Lower body negative pressure (LBNP) is not a true orthostatic stress, because the body is not upright and does not experience a true G force in the $+G_z$ direction. However, LBNP is a simulation of orthostatic stress that enables investigators to keep their subjects passive and relaxed, while suction on the subjects' lower bodies elicits fluid shifts similar to those that occur during orthostatic stress. One major difference between true orthostatic stress and LBNP are the hydrostatic pressure gradients between the heart and brain and between the heart and carotid sinuses that exist with orthostasis but not with LBNP. Thus receptors in these regions may experience differing stimulation when the subjects respond to LBNP than they do when exposed to true orthostasis, and they may initiate differing reflex responses.

1. Responses and Comparisons with Men's Responses

Several investigators have examined the responses of women to LBNP, and in some instances they have used the same protocol to test both men and women.

TABLE 6
Percentage Change (Mean ± SEM) in Circulatory Responses
During 70° Tilt in Younger (20 to 29 years) and Older
(40 to 49 years) Sedentary and Fit Women

	Young fit (n=8)	Young sedentary (n=8)	Older fit (n=8)	Older sedentary (n=8)
Heart rate	38.2 ± 6.7	42.1 ± 6.0	32.3 ± 6.2	28.9 ± 4.5
Stroke volume	−40.5 ± 4.0	−45.0 ± 4.5	−40.0 ± 4.3	−39.4 ± 4.8
Cardiac index	−18.4 ± 5.2	−23.2 ± 4.8	−22.1 ± 3.3	−22.1 ± 3.3
Transthoracic impedance	11.4 ± 1.1	9.5 ± 0.7	12.0 ± 1.0	9.9 ± 1.1
Peripheral vascular resistance index	37.3 ± 12.1	41.3 ± 9.1	39.7 ± 6.8	45.2 ± 12.8
Systolic pressure	−5.6 ± 2.1	−5.0 ± 3.7	−3.6 ± 1.5	−0.0 ± 3.1
Diastolic pressure	19.4 ± 2.5	14.4 ± 2.8	16.2 ± 2.3	13.9 ± 3.7
Pulse pressure	−37.9 ± 6.1	−34.5 ± 9.2	−37.9 ± 6.1	−21.3 ± 9.0

Data from Porth, C. M., Barney, J. A., Hughes, C. V., Ptacin, M. J., Sheldahl, L., and Smith, J. J., in preparation.

Frey et al. exposed 20 women (mean age = 33.3 years) to a protocol that used graduated decreases in pressure within the LBNP device: 0, 8, 16, 30, 40, and 50 mmHg below atmospheric pressure.[61] All measured cardiovascular variables of these women changed significantly between readings at 0 mmHg and those at −50 mmHg, except diastolic blood pressure, which tended to increase. Values at rest in the LBNP device with no suction and at −50 mmHg LBNP are listed in Table 7. These responses are similar to those (reported by another laboratory) for men,[62] except that these women had greater increases in heart rate during LBNP than did the men.

In another report, Frey and Hoffler compared the responses to LBNP of 21 women (35.3 years of age; 36.3 ml/kg/min maximum oxygen uptake) with those of 26 men (36.2 years of age; 37.3 ml/kg/min maximum oxygen uptake).[63] All variables (the same variables as those reported in the study referenced above[61]) except mean arterial pressure were changed with LBNP. The only differences between the sexes were that the men had higher systolic and higher mean arterial pressures than the women before the application of negative pressure and at all levels of LBNP. The magnitude of responses in the blood pressure to LBNP did not differ between the sexes, however. During LBNP, men experienced a greater increase in leg circumference (fluid accumulation) and total peripheral resistance; women experienced greater increases in thoracic impedance (greater reduction in thoracic fluid) and in heart rate ($p < 0.05$ for all variables except heart rate where $p < 0.1$). The authors suggested that the men had more leg muscle than the women had; therefore, they had more vascularity and interstitial space in which to sequester fluid. However, the women apparently experienced a greater decrease in thoracic fluid than the

TABLE 7

Mean Values for Cardiovascular and Fluid Variables of Women (n = 20) at Rest and During −50 mmHg LBNP

Variable	0 mmHg	−50 mmHg
Heart rate, bpm	66	91
Stroke volume, ml	79	40
Cardiac output, l/min	5.0	3.4
Systolic blood pressure, mmHg	101	92
Diastolic blood pressure, mmHg	68	69
Mean arterial pressure, mmHg	78	76
Thoracic impedance, Ω	29	32
Ejection time, ms	319	228
Change in leg circumference, %	—	1.0

Data from Frey, M. A. B., Mathes, K. L., and Hoffler, G. W., *Aviat. Space Environ. Med.*, 57, 531, 1986.

men did and thus must have been accumulating fluid in other regions of the lower body. None of these subjects became presyncopal and, despite the sex differences in the responses in some cardiovascular variables, both sexes maintained their arterial pressures. However, the sex differences observed in this study for LBNP agree only in part with those described above for standing.[7] In both studies, the women had greater changes than the men did in the heart rate and thoracic impedance, and the changes in arterial pressure were similar for the women and the men. However, in this study of LBNP the men and the women had similar stroke volumes and the men had greater increases in peripheral vascular resistance with LBNP than did the women. This difference between studies may be related to the low minimum oxygen uptake values of the men in this study.

Hordinsky et al. studied responses to LBNP of 12 men (30 to 42 years of age, 38.0 ml/kg/min maximum oxygen uptake) and 12 women (24 to 38 years of age, 33.8 ml/kg/min maximum oxygen uptake) before and after 6 h of bedrest or water immersion deconditioning.[64] The investigators progressively lowered pressure within the LBNP device until the subjects became presyncopal. The women had higher heart rates (80 vs. 71 bpm) and lower systolic blood pressures (116 vs. 125 mmHg) than the men had. The only significant differences between the sexes during LBNP before deconditioning was a greater accumulation of fluid in the legs of the men. These data are consistent with those of Frey et al.[63] The women's LBNP tolerance time before deconditioning was slightly less than the men's but was not significantly different.

Goldwater et al.[65] also studied men and women before and after deconditioning (10 d bedrest) and also reported that, before the bedrest period, LBNP tolerance times did not differ significantly between men and women. However, the men's tolerance times were slightly longer (14.38 vs. 12.88 min), and four of the nine women, but only one of the eight men, became presyncopal during 15 min of −50 mmHg LBNP.

Montgomery et al. subjected six men (mean age = 25 years) and four women (mean age = 23.5 years) to repeated tests of 5 min each of rapid-onset LBNP at −20, −40, and −60 mmHg presented in random order.[66] The women had higher heart rates than the men had at all LBNP pressures, and the men had greater systolic blood pressures at −40 and −60 mmHg and greater increases in leg volume than the women had at all pressures. The women reported more discomfort during LBNP at −40 and −60 mmHg and were unable to complete 10 of the 12 tests at −60 mmHg. The men completed all the tests. The authors suggested that the lower tolerance of their female subjects might have resulted from greater apprehension, obstruction of venous return during LBNP by the uterus, or lesser fitness and muscle mass, as compared with the male subjects.

In summary, differences between men and women in several of the cardiovascular responses to LBNP are consistently observed: women have higher heart rates and greater reductions in thoracic fluid; men have higher systolic blood pressures and greater increases in leg size. Some data suggest that men have greater tolerance to this stress; however, most data indicate that, although their response mechanisms differ, women and men have similar tolerance.

2. Effects of Menstrual Cycle Phase

In the study of women's responses to LBNP (Frey et al.) discussed above, each of the 20 women were tested on two occasions in random order: once between days 3 and 10 of the menstrual cycle (follicular phase) and once between days 21 and 25 of the menstrual cycle (luteal phase).[61] The menstrual phase and the ovulatory nature of the women's cycles were verified by analyses of blood hormone levels. Follicle-stimulating hormone was 7.7 ± 3.2 mIU/ml in the follicular phase and 4.6 ± 2.4 mIU/ml in the luteal phase ($p <0.01$). Progesterone was 19 ± 15 µg/dl in the follicular phase and 1101 ± 363 µg/dl in the luteal phase ($p <0.01$). The following variables did not differ between the phases of the menstrual cycle when subjects were resting supine or when they were responding to LBNP: weight, hemoglobin, hematocrit, leg size, or any of the cardiovascular variables listed above in Table 7. Four subjects became presyncopal during LBNP in the follicular phase of their menstrual cycles and one subject in the luteal phase, which was not a significant difference by the chi-square test. The authors suggested that, although hormone assays verified that the subjects were in the luteal phase of their menstrual cycles when they were tested between days 21 and 25, the peak increase in plasma volume, which might have contributed to increased orthostatic tolerance or to modulation of responses to LBNP, might not have occurred until several days later.

Twenty-three women were tested with LBNP at the Ames Research Center before and after a 2-week bedrest period without regard to the phase of the menstrual cycle, and no differences in tolerance or responses to LBNP were detected as a function of the time in the menstrual cycle.[67]

3. Effects of Fitness

Frey et al. studied 45 women who ranged in aerobic capacity from 23 to 55.3 ml/kg/min (mean = 37.8 ± 6.71 ml/kg/min). The subjects' body weight was 28.3% fat (determined from skinfold measurements), and this was negatively correlated with their aerobic capacity.[68] Each subject had a progressive LBNP test to –50 mmHg and a treadmill test (using a modified Bruce protocol) to peak oxygen uptake. The variables that were measured before and during LBNP are those listed above in Table 7. During LBNP, all measured variables changed, except diastolic and mean arterial pressures: heart rate increased 34%, stroke volume decreased 48%, total peripheral resistance increased 57%, thoracic impedance increased 9%, and calf circumference increased 1%. Systolic and mean arterial pressures were the only variables, of those measured while the subjects were at rest before they were exposed to LBNP, that were correlated with aerobic capacity; they were directly correlated. Six subjects experienced signs or symptoms of impending syncope during the LBNP tests. Their mean peak aerobic capacity was 39.2 ml/kg/min, not significantly different ($p >0.6$) from the mean peak aerobic capacity (37.6 ml/kg/min) of the subjects who did not become presyncopal. The only significant correlation between peak aerobic capacity and the response to LBNP was a positive correlation with percent change in calf circumference at –30 and –40 mmHg LBNP (but not at –50 mmHg). This indicates more fluid pools in the legs of fit subjects during LBNP; however, other mechanisms must have compensated for this difference, since no difference existed in the arterial pressure response to LBNP. The explanation for the greater pooling of blood in the legs of the more fit women as compared with the less-fit women may be the same as for the greater pooling of blood in the legs of men as compared with women — they have a greater percentage of muscle and, therefore, more vascularity and interstitial space for sequestration.

Hudson et al. compared the responses of eight physically "trained" women (23.3 years of age,

56.8 ml/kg/min maximum oxygen uptake) and eight "untrained" women (25.9 years of age, 39.4 ml/kg/min maximum oxygen uptake) to a graduated protocol of LBNP to –50 mmHg.[69] These investigators did not control for the stage of the menstrual cycle, but they performed no tests during the subjects' menses. In addition to having greater maximum oxygen uptake, the trained subjects had greater plasma volume per kilogram and blood volume per kilogram and lesser body weight, surface area, and leg volume. Heart rate, cardiac index (CO_2 rebreathing), and forearm blood flow did not differ between subject groups. Responses to LBNP were as expected: heart rate, peripheral vascular resistance, forearm vascular resistance, and leg volume increased; cardiac index, stroke index, systolic blood pressure, and mean arterial pressure decreased; and diastolic blood pressure was not changed. No subjects in either group had symptoms or signs of impending syncope, and groups did not differ in their responses to LBNP for cardiac index, stroke index, systolic blood pressure, heart rate, forearm vascular resistance, or the ratio of change in heart rate to change in systolic blood pressure, which those authors considered to be a measure of arterial baroreceptor responsiveness. However, the group of untrained women had a greater peripheral vascular resistance response at –50 mmHg LBNP and a greater change in leg volume at pressures from –8 to –32 mmHg than the group of trained women. This contradicts data from the study by Frey et al.[68] We have no explanation for this difference. The LBNP protocols were similar, but the subjects studied by Frey et al. were older.

Torikoshi et al. studied the responses of eight women (mean age = 22 years; 41.8 ml/kg/min maximum oxygen uptake) to a graduated protocol of LBNP to –60 mmHg and observed a significant positive correlation of maximum oxygen uptake with both LBNP tolerance pressure and LBNP tolerance time.[70]

These three studies indicate that for women tolerance and responses to orthostatic or orthostaticlike stresses are independent of, or positively associated with, aerobic capacity.

4. Effects of a Preceding Bedrest, Head-Down Tilt, or Water Immersion

As introduced previously, 12 men and 12 women were subjected by Hordinsky et al. to LBNP tests before and after two types of "deconditioning": 6 h bedrest and 6 h water immersion to the neck.[64] Each subject performed two tests. The only sex differences in the responses to LBNP before deconditioning were a greater pooling of blood in the legs of men and a slightly lesser, though not significant, tolerance time to LBNP for the women. The women's tolerance time was decreased after both bedrest and water immersion, but the men's tolerance time was decreased only after water immersion. The authors also noted that a long tolerance time coincided with a large increase in plasma renin activity, especially in the men.

Several studies have been performed on female subjects in the bedrest facility at the NASA Ames Research Center. In the first study, which was done in 1973, eight women (23 to 34 years) were tested with LBNP at –50 mmHg for 15 minutes, or until signs or symptoms of impending syncope occurred, before and after 17 d of bedrest.[67] All subjects completed the 15-min test before bedrest; none completed it after bedrest. After bedrest, subjects' heart rates increased more during LBNP than they had before bedrest (102 vs. 136 bpm after 10 min of LBNP), and the tolerance time to LBNP was significantly reduced. Sandler and Winter stated that after 2 weeks of bedrest women had a greater loss of tolerance time to LBNP and a greater heart rate response than men had.[71]

The following data are from other studies that were performed at the Ames Research Center using the same protocol. Montgomery et al. used an impedance technique to measure blood volumes in several regions of the lower body of ten women (34 to 45 years) during LBNP before and after 9 d of bedrest.[72] Blood flow in the legs was decreased after bedrest in absolute amount, but not in ml/min/100 g. The response of leg blood flow to LBNP did not differ from before to after bedrest. Blood flow in the pelvic area did not change after bedrest, but less blood pooled in the pelvic area during LBNP after bedrest.

Goldwater et al. reported that LBNP tolerance time was decreased after 9 d of bedrest (by

10.8%), not a significant change, and that LBNP tolerance time was not correlated with maximum oxygen uptake.[73] Polese et al. reported that, in a group of eight women (40 to 55 years), left ventricular end-diastolic volume measured with echocardiography decreased from 87.6 to 83.7 ml after 10 d of bedrest.[74] Heart rate increased more and isovolumetric contraction time decreased more during LBNP after bedrest than they had before bedrest. In a study (Goldwater et al.) in which eight men and nine women (55 to 65 years) underwent 10 d of bedrest, the women lost more plasma volume and had shorter tolerance times to LBNP after bedrest than the men (7.97 min vs. 14.64 min); they had a greater fall in systolic blood pressure and a lesser increase in heart rate during LBNP.[65] The authors suggested that these differences may have be due to the smaller ventricular volume of the women. They suggested that the left ventricle shrinks during LBNP; then when heart rate increases, filling is further curtailed. Thus, mechanoreceptors in the left ventricle are stimulated, and a reflex that is initiated via unmyelinated vagal afferent fibers induces syncope.

These data indicate rather convincingly that short periods of bedrest, from hours to several weeks, reduce the ability of women to withstand orthostatic or orthostaticlike stresses. Apparently the orthostatic tolerance of women is affected more by these short periods of bedrest than is that of men. We have no data indicating how well interventions, such as ingestion of fluid before arising, counteract this loss of tolerance in women.

D. CENTRIFUGATION (+G_z)

Knowledge of human responses to gravitational forces greater than those at the Earth's surface (1 G) is important to the National Aeronautics and Space Administration at levels of approximately two G_z, because astronauts encounter these forces when returning to earth in the space shuttle. Space travelers are particularly sensitive to these forces, because their blood volume is reduced after adaptation to the weightless environment of space. Human responses to +G_z stresses of a greater magnitude, up to 10 or 12 G, are important to the Air Force, because pilots of high-performance fighter aircraft encounter these forces and are at risk of blacking out. Experiments to test +G_z tolerance are usually performed in a centrifuge. Unlike LBNP, centrifugation elicits a "hydrostaticlike" pressure gradient between the heart and the brain. Women now perform as both astronauts and flight-qualified pilots, and a few studies of their responses and tolerances to +G_z have been performed.

1. Responses and Comparisons with Men's Responses

Gillingham et al. studied the responses of 102 women (19 to 41 years) to a series of tests in the centrifuge.[75] In the first test, the G_z forces were increased gradually until the subjects reached a visual "endpoint" of 100% loss of peripheral vision or 50% loss of central vision, which were carefully monitored by established criteria. This test was followed by a series of tests with a rapid onset of G forces, each continuing 15 s or until a visual endpoint. The final test was another gradual-onset run. Data from these women were compared with data from 128 men who had been tested previously with the same protocol. Seventy-seven women and 82 men completed all profiles. Tolerance scores of the women and those of the men were equal. G_z tolerance was determined to be inversely correlated with height ($p < 0.001$) and directly correlated with weight ($p < 0.05$); it was independent of age and only slighted affected by the physical activity level. After the men's and women's G_z tolerance times were corrected for the effects of age, height, weight, and physical activity, the effect of sex ($p \cong 0.1$) indicated that women might have a lesser tolerance if they were the same height, weight, age, and physical activity as men. Since they are usually shorter, their tolerance tends to be better than that of men. Eighty-eight percent of the women and 80% of the men completed 15 s of 7 G in the training protocol. Forty-seven percent of the women and 49% of the men had symptoms; 53% of the women and 56% of the men had dysrhythmias. The women reported apprehension more frequently than men did. The authors concluded that the G tolerances of women and men are equal for up to 7 G for 15 s.

2. Effects of Menstrual Cycle Phase

In the above-described study, menstrual-cycle data were obtained from 47 subjects who were tested twice — once during menses and once during some other stage of their cycles.[75] No differences existed in responses to centrifugation between the tests performed during menses and those performed during other times in the menstrual cycle.

3. Effects of Fitness and Heat Acclimation

Greenleaf et al. and Berry et al. have conducted several investigations to determine whether exercise training in the heat affects tolerance to $+G_z$.[76,77] Fifteen women were tested for $+G_z$ tolerance before and after 8 d during which they either exercised 2 h per day in the heat (40.6°C), exercised 2 h per day in the cool (18.7°C), or were sedentary in the cool.[76] Tolerance to $+G_z$ was not changed after the exercise training period in any of the groups.

In another study, four men (27 years of age, 64 ml/kg/min maximum oxygen uptake) and six women (22 years of age, 47 ml/kg/min maximum oxygen uptake) performed ergometer exercise 2 h a day in the heat (40°C, 42% relative humidity) for 12 d.[60] Tolerance to $+G_z$ stress was not changed in either sex. After the heat-acclimation period, however, the men's heart rate during centrifugation was lower than it had been before the heat acclimation period.

4. Effects of a Preceding Bedrest, Head-down Tilt, or Water Immersion

Eight women (23 to 34 years) were subjected to $+3$ G_z stress three times with 5-min rests between, before, and after spending 14 d in bed at the Ames Research Center.[78,79] Tolerance time to the $+G_z$ stress was decreased from 388 s before the bedrest period to 198 s after bedrest. This was a greater loss of tolerance than experienced by a group of young men who had been tested previously. Plasma volume of the women was reduced 12.6% during centrifugation before bedrest but was reduced only 4% during centrifugation after bedrest. The correlation between the decrease in plasma volume and the decrease in tolerance to centrifugation was 0.72, indicating that more than half the loss of tolerance resulted from factors other than the loss of plasma volume.[79] Plasma renin activity of these subjects when they were at rest was greater after bedrest than it was before bedrest but it was not affected by centrifugation.[79] Examination of data acquired previously from a group of men indicated that the plasma renin activity of the men also increased after bedrest, but the men's plasma renin activity was increased after 20 to 30 min of $+2.5$ G_z. The authors suggested that the shorter tolerance time of the women might have prevented plasma renin activity from increasing a significant amount during centrifugation. Arginine vasopressin of the women was decreased 33% after bedrest, but this was not significantly different from values before bedrest. However, arginine vasopressin of the women increased significantly during centrifugation both before and after the bedrest period, to a greater extent than it increased during centrifugation in the men.

In another study at the Ames Research Center, the $+G_z$ tolerance of ten women (35 to 44 years) was compared with that of seven men in the same age range who had been tested previously.[80] Before bedrest, the women had lower $+G_z$ tolerance than the men. The women's heart rates increased from 75 (at rest) to 109 bpm with $+2$ G_z centrifugation and from 75 to 144 bpm with $+3$ G_z. After bedrest, the women's heart rates increased from 80 to 134 bpm with $+2$ G_z centrifugation and from 82 to 160 bpm with $+3$ G_z. The women's tolerance to $+2$ G_z decreased to 65% of prebedrest tolerance time and their tolerance to $+3$ G_z decreased to 43% of the prebedrest tolerance time. The men had no loss of tolerance to $+2$ G_z, and their tolerance decreased only to 50% of prebedrest levels with $+3$ G_z centrifugation.

Eight men and nine women (55 to 65 years) were tested at $+1.5$, $+2$, and $+3$ G_z before and after 10 d of bedrest.[65] Tolerance times did not differ between the sexes before bedrest, but the women lost more tolerance to $+G_z$ as a result of the 10 d bedrest period.

These data reinforce the data of women's and men's responses to LBNP before and after bedrest and indicate that, while tolerance to these stresses usually does not differ between normal active men and normal active women, women tend to lose more of their tolerance to orthostaticlike stresses after a period of bedrest.

VI. CONCLUSIONS

This review of responses of women and men to orthostatic, orthostaticlike, and nonorthostatic stresses has revealed that, while response patterns and tolerances appear to be generally similar, some differences exist between the responses of women and those of men. Most of these differences in responses are consistent from research study to research study, and even from stress to stress. In fact, women are reported to respond to most stresses studied with a greater tachycardia than men do, and the men studied responded with greater blood pressure elevation, especially systolic blood pressure. Furthermore, men generally have shown greater responsiveness to catecholamines during stress.

During exercise, the hearts of the men studied responded with greater increases in stroke volume, which appeared to be the result of greater contractility (decreased end-systolic volume), and, unlike the hearts of the women studied, with increased ejection fraction.

During LBNP, the legs of the men increased more in size, indicating more fluid was pooling there. However, the women examined lost more fluid from the thorax (their thoracic impedance increased more), indicating that fluid was pooling at a site in their lower bodies that still awaits discovery. After a period of deconditioning, women were shown to lose more of their tolerance to orthostaticlike stresses than the men did. None of the studies reported indicated that aerobic fitness had a deleterious effect on the orthostatic responses of the women; this is still an area of controversy in regards to men.

With aging, men have been shown to lose more of their isometric strength and endurance, as well as more stroke volume, ejection fraction, and cardiac output than women.

We believe that the studies we have reviewed merely hint at the similarities and differences in women's and men's responses to stress and that more data are needed before these similarities and differences can be well explained. However, in the following discussion we will summarize the known differences between the sexes that may provide some explanations.

Differences in hormones, especially the reproductive hormones, provide one possible explanation. This, in fact, may explain the differences in body size, muscle mass, fat distribution, and possibly heart size, hematocrit, and the hemoglobin content of blood. The larger vital capacity of men is directly related to their greater body size and height. The greater sequestration of blood in the legs of men during LBNP may result from their larger leg muscle mass; however, this does not explain where the fluid is sequestered in women. Smaller chest size, smaller heart size, and greater fat-to-lean ratio of tissue between thoracic impedance electrodes may explain the greater thoracic impedance of women. The higher hemoglobin concentration and hematocrit of men, plus their larger blood volume, provides a greater oxygen-carrying capacity for their blood and a greater margin for increasing the arterial-venous oxygen difference. Women meet their metabolic needs through higher heart rates and greater heart-rate increases.

Differences in blood catecholamine levels and in autonomic nervous system function may cause some of the responses to exercise and to nonexercise stresses to be mediated differently in women than they are in men. Men reportedly secrete more catecholamines during mental and other nonexercise stresses, whereas women secrete more catecholamines during exercise. The greater responsiveness of men's vessels to α- and β-adrenergic stimuli and the greater ability of their hearts to increase contractility and stroke volume may explain in part their greater blood pressure responses to stress and their better ability to withstand orthostatic stress under some circumstances. More advanced atherosclerosis in the vessels of men would also contribute to higher blood pressures at rest and during stress. Men also have higher total blood cholesterol levels and lower high-density lipoprotein cholesterol levels than women have. For men, these are acknowledged risk factors for CAD that may contribute to their blood pressure elevation with aging by increasing peripheral vascular resistance and decreasing arterial compliance.

In attempting to explain sexual differences in response to stress, however, we must not confuse physiological causes with psychological or environmental causes. Some reported differences may result from differences between the social or professional status of women and those of men. To interpret each set of data that compares responses of women with those of men, certain questions should be asked: (1) Did the women and the men in the study have similar activities and daily environments? (2) Did the women and the men in the study have similar expectations for themselves and were others' expectations for the women and for the men similar? (3) Did the women and the men in the study have similar levels of daily exercise and changes in levels of exercise with respect to age? and (4) As future studies are conducted, will some of the reported physiological and pathological differences disappear as women and men share more of the same roles?

A. DIRECTIONS FOR FUTURE RESEARCH

This chapter is intended as a starting point in the process of identifying and understanding similarities and differences between orthostatic function of women and men for the improvement of human health and function. The literature presented here has revealed some wide gaps in the current knowledge about the responses of women to orthostatic stress and about comparisons of women's and men's responses to orthostatic stress. Some research needs are very specific. Studies are needed comparing women's and men's responses to head-up tilt, a true orthostatic stress. These studies should include all or many of the variables discussed above. Data are particularly scarce and are needed for responses in cerebral blood flow; arterial baroreceptor reflex responsiveness; and hormone levels such as renin, aldosterone, antidiuretic hormone, and atrial natriuretic factor. Where does the fluid go in women during orthostatic stress or LBNP? New Doppler and impedance techniques should allow researchers to gain this information noninvasively. The answers to these questions might also help us in understanding human responses to orthostatic stress and help us develop interventions to improve function or to decrease disease.

Research should be designed to optimize the understanding of physiological differences between women and men and between their responses to stress. The health-care professional should take advantage of this information to improve the health and well-being of both women and men. Studies should be designed for adequate numbers of both female and male subjects, using the same protocols for both, but not combining data unless, and until, the groups have been compared and comparability between them is practically certain.

Actual and factual differences, or similarities, throughout the menstrual cycle for cardiovascular, hormonal, and fluid variables should be carefully documented.

Finally, studies to reveal the mechanisms of differences between women and men in normal function and in responses to stress should be designed and implemented.

REFERENCES

1. **Kannel, W. B., Hjoreland, M. C., McNamara, P. M., Gordon, T.,** Menopause and risk of cardiovascular disease, *Ann. Intern. Med.,* 85, 447, 1976.
2. **Katch, F. I. and Katch, V. L.,** Optimal health and body composition, in *Women and Exercise: Physiology and Sports Medicine,* Shangold, M. and Mirkin, G., Eds., F. A. Davis, Philadelphia, 1988, 30.
3. **Astrand, I.,** Aerobic work capacity in men and women with special reference to age, *Acta Physiol. Scand.,* Suppl 169, 92, 1960.
4. **Astrand, P. O.,** Human physical fitness with special reference to sex and age, *Physiol. Rev.,* 36, 307, 1956.

5. **Gillum, R. F.,** The epidemiology of resting heart rate in a national sample of men and women: associations with hypertension, coronary heart disease, blood pressure, and other cardiovascular risk factors. I, *Am. Heart J.,* 116 (1), 163, 1988.

6. **Montoye, J., Willis, P. W., III, Howard, G. E., and Keller, J. B.,** Cardiac preejection period: age and sex comparisons, *J. Gerontol.,* 26, 208, 1971.

7. **Frey, M. A. B., Tomaselli, C. M., and Freeman, M.,** Cardiovascular responses to postural change: differences by sex and age, submitted.

8. **Frey, M. A. B, Doerr, B. M., and Miles, D. S.,** Transthoracic impedance: differences between men and women with implications for impedance cardiography, *Aviat. Space Environ. Med.,* 53, 1190, 1982.

9. **McKinney, M. E., Buell, J. C., and Eliot, R. S.,** Sex differences in transthoracic impedance: evaluation of effects on calculated stroke volume index, *Aviat. Space Environ. Med.,* 55, 893, 1984.

10. **Porth, C. M., Barney, J. A., Hughes, C. V., Ptacin, M. J., Sheldahl, L., and Smith, J. J.,** The effect of age and physical fitness on graded postural stress in women, in preparation.

11. **Freedman, R. R., Subhash, C., Sabharwal, S. C., and Desai, N.,** Sex differences in peripheral vascular adrenergic receptors, *Circ. Res.,* 61, 581, 1987.

12. **Kelleher, C., Joyce, C., Kelly, G., and Ferriss, J. B.,** Blood pressure during the normal menstrual cycle, *Br. J. Obstet. Gynaecol.,* 93, 523, 1986.

13. **Greenberg. G., Imeson, J. D., Thompson, S. G., and Meade, T. W.,** Blood pressure and the menstrual cycle, *Br. J. Obstet. Gynaecol.,* 92, 1010, 1985.

14. **Engel, P. and Hildebrandt, G.,** Rhythmic variations in reaction time, heart rate and blood pressure at different times in the menstrual cycle, in *Biorhythms and Human Reproduction,* Ferien, M., Halberg, F., Richart, R. M., and Vande Wiele, R. L., Eds., Wiley-Interscience, New York, 1973, 325.

15. **Freedman, S. M., Ramcharan, S., and Hoag, E.,** Some physiological and biochemical measurements over the menstrual cycle, in *Biorhythms and Human Reproduction,* Ferien, M., Halberg, F., Richart, R. M., and Vande Wiele, R.L,.Eds., Wiley-Interscience, New York 1973, 259.

16. **James, G.D., Yee, L.S., Harshfield, G.A., and Pickering, T.G.,** Sex differences in factors affecting daily variation of blood pressure, *Soc. Sci. Med.,* 26, 1019, 1988.

17. **Hughes, C. J., Asmar, R. G., London, G. M., and Safar, M. E.,** Age- and sex-related changes in ratio between ankle and brachial systolic pressure changes in normal subjects, *Angiology,* 39, 219, 1988.

18. **Daniels, S. R., Heiss, G., Davis, C. E., Hames, C. G., and Tyroler, H. A.,** Race and sex differences in correlates of blood pressure change, *Hypertension,* 11, 249, 1988.

19. **Kannel, W. B. and Abbott, R. D.,** Incidence and prognosis of myocardial disease in women: the Framingham Study, in *Coronary Heart Disease in Women: A Summary of the Proceedings of an NIH Workshop,* Eaker, E. D., Packard, B., Wenger, N. K., Clarkson, T. B., and Tyroler, H. A., Eds., National Heart, Lung, Blood Institute, Bethesda, MD., 1986, 208.

20. **Eaker, E. D., Packard, B., Wenger, N. K., Clarkson, T. B., and Tyroler, H. A., Eds.,** *Coronary Heart Disease in Women: A Summary of the Proceedings of an NIH Workshop,* National Heart, Lung, Blood Institute, Bethesda, 1986.

21. **Wenger, N. K. and Roberts. P.,** Session III highlights: clinical aspects of coronary heart disease in women, in *Coronary Heart Disease in Women: A Summary of the Proceedings of an NIH Workshop,* Eaker, E. D., Packard, B., Wenger, N. K., Clarkson, T. B., and Tyroler, H. A., Eds., Bethesda, MD., NHLBI, 1986, 22.

22. **Boucek, R. J., Romanelli, R., Willis, W. H., Jr, and Mitchell, W. A.,** Sex differences in obstructive coronary artery disease in patients 65 years of age or older with angina pectoris, *Circulation,* 66, 926, 1982.

23. **Mock, M. B., Rinqvist, I., Fisher, L. D., Davis, K. B., Chaitman, B. R., Kouchoukos, N. T., Kaiser, G. C., Alderman, E., Ryan, T. J., Russell, R. O., Mullins, S., Fray, D., Killip, T., and Participants in the Coronary Artery Surgery Study,** Survival of medically treated patients in the Coronary Artery Surgery Study (CASS) registry, *Circulation,* 66, 562, 1982.

24. **Schatzkin, A., Cupples, L. A., Heeren, T., Morelock, S., and Kannel, W. B.,** The epidemiology of sudden unexpected death: risk factors for men and women in the Framingham Heart study, *Am. Heart J.,* 107, 1300, 1984.

25. **Coleman, R. W.,** The role of platelets in the genesis of ischemia, in *Sudden Coronary Death,* Vol. 32, Greenberg, H. M. and Dwyer E. M., Eds., Annals of the New York Academy of Science, New York, 1982, 190.

26. **Prentice, R. L., Szatrowski, T. P., and Fukikura, T.,** Leukocyte counts and coronary heart disease in a Japanese cohort, *Am. J. Epidemiol.,* 116, 496, 1982.

27. **Detry, J. M. R., Kapita, B. M., Cosyns, J., Sottiaux, B., Brasseur, L. A., and Rousseau, M. F.,** Diagnostic value of history and maximal exercise electrocardiography in men and women suspected of coronary heart disease, *Circulation,* 56, 756, 1977.

28. **Sketch, M. H., Mohiuddin, S. M., Lynch, J. D., Zenka, A. E., and Runco, V.,** Significant sex differences in the correlation of electrocardiographic exercise testing and coronary arteriograms, *Am. J. Cardiol.* 36, 169, 1975.

29. **Kusuml, F. and Bruce, R. A.,** Unpublished observations; cited in **Bruce, R.A.,** Values and limitations in exercise electrocardiography, 50, 1, 1974.

30. **Nakhjavan, E. K., Natarajan, G., Seshachary, P., and Golderberg, H.,** The relationship between prolapsing mitral leaflet syndrome and angina and normal coronary arteriograms, *Chest,* 70, 706, 1976.

31. **Vogel, J. A., Patton, J. F., Mello, R. P., and Daniels, W. L.,** An analysis of aerobic capacity in a large United States population, *J. Appl. Physiol.,* 60, 494, 1986.

32. **Astrand, I., Astrand, P. O., Hallback, I., and Kilbom, A.,** Reduction in maximum oxygen intake with age, *J. Appl. Physiol.,* 35, 649, 1973.

33. **Drinkwater, B.,** Physiological responses of women to exercise, in *Exercise and Sport Sciences Reviews,* Vol. 10, Wilmore, J., Ed., Academic Press, New York, 1973, 125.

34. **Sparling, P.,** A meta-analysis of studies comparing maximal oxygen uptake in men and women, *Res. Q. Exerc. Sport,* 51, 542, 1980.

35. **Astrand, P. O., Cuddy, T. E., Saltin, B., and Stenberg, J.,** Cardiac output during submaximal and maximal work, *J. Appl. Physiol.,* 19, 268, 1964.

36. **Hossack, K. F. and Bruce, R. A.,** Maximal cardiac function in sedentary normal men and women —comparison of age-related changes, *J. Appl. Physiol.,* 53, 799, 1982.

37. **Higginbotham, M. B., Morris, K. G., Coleman, R. E., and Cobb, F. R.,** Sex-related differences in the normal cardiac response to upright exercise, *Circulation,* 70, 357, 1984.

38. **Adams, F. K., Vincent, L. M., McAllister, S. M., El-Ashmawy, H., and Sheps, D. S.,** The influence of age and gender on left ventricular response to supine exercise in asymptomatic normal subjects, *Am. Heart J.,* 113, 732, 1987.

39. **Cureton, K., Bishop, P., Hutchinson, P., Newland, H., Vickery, S., and Zwiren, L.,** Sex difference in maximal oxygen uptake, *Eur. J. Appl. Physiol.,* 54, 656, 1986.

40. **Convertino, V. A., Stremel, R. W., Bernauer, E. M., and Greenleaf, J. R.,** Cardiorespiratory responses to exercise after bedrest in men and women, *Acta Astronautica,* 4, 895, 1977.

41. **Convertino, V. A., Goldwater, D. J., and Sandler, H.,** Bedrest-induced peak Vo_2 reduction associated with age, gender, and aerobic capacity, *Aviat. Space Environ. Med.,* 57, 17, 1986.

42. **Wagner, J. A. and Horvath, S. M.,** Cardiovascular reactions to cold exposures differ with age and gender, *J. Appl. Physiol.,* 58, 187, 1985.

43. **Frey, M.A.B, Siervogel, R. M., Selm, E. A., and Kezdi, P.,** Cardiovascular responses to cooling of limbs determined by noninvasive methods, *Eur. J. Appl. Physiol.,* 44, 67, 1980.

44. **LeBlanc, J., Cote, J., Dulac, S., and Dulong-Turcot, F.,** Effects of age, sex, and physical fitness on responses to local cooling, *J. Appl. Physiol.,* 44, 813, 1978.

45. **Hastrup, J. L. and Light, K. C.,** Sex differences in cardiovascular stress responses: modulation as a function of menstrual cycle phases, *J. Psychosom. Res.,* 28, 475, 1984.

46. **Mannino, J. A. and Washburn, R. A.,** Cardiovascular responses to moderate facial cooling in men and women, *Aviat. Space and Environ. Med.,* 58, 29, 1987.

47. **Petrofsky, J. S., Burse, R. L., and Lind, A. R.,** Comparison of physiological responses of women and men to isometric exercise, *J. Appl. Physiol.,* 38, 863, 1975.

48. **Kobryn, U., Hoffman, B., and Ransch, B.,** Sex- and age-related blood pressure responses to dynamic work with small muscle masses, *Eur. J. Appl. Physiol.,* 44, 813, 1978.

49. **Booth, R. W., Ryan, J. M., Mellett, H. C, Swiss, E., and Neth, E.,** Hemodynamic changes associated with the Valsalva maneuver in normal men and women, *J. Lab. Clin. Med.,* 59, 275, 1962.

50. **Frey, M. A. B., Bloom, H. R., and Miles, D. S.,** Cardiovascular changes during graded mental stress, *Fed. Proc.,* 46, 668, 1987.

51. **Forsman, L. and Lindblat, L. E.,** Effect of mental stress on baroreceptor-mediated changes in blood pressure and heart rate and on plasma catecholamines and subjective responses in healthy men and women, *Psychosom. Med.,* 45, 435, 1983.

52. **Frankenhouser, M.,** The sympathetic-adrenal and pituitary response to challenge, in *Behavioral Basis of Coronary Heart Disease,* Dembroski, T. M., Schmidt, T. H., and Bumchen, G., Eds., Karger, Basel, 1983, 91.

53. **Turner, A.,** The circulatory minute volumes of healthy young women in reclining, sitting, and standing positions, *Am. J. Physiol.,* 80, 601, 1927.

54. **Turner, A.,** The adjustment of heart rate and arterial blood pressure in healthy young women during prolonged standing, *Am. J. Physiol.,* 81, 197, 1927.

55. **Frey, M. A. B., Doerr, B. M., Mann, B. L., and Miles, D. S.,** Comparison of impedance ventricular function indices with systolic time intervals, *Proc. IEEE NAECON,* 1086, 1981.

56. **Moore, K. I. and Newton, K.,** Orthostatic heart rates and blood pressures in young healthy women and men, *Heart Lung,* 15, 611, 1986.

57. **Turner, A. H., Newton, I., and Haynes, F. W.,** The circulatory reaction to gravity in healthy young women, *Am. J. Physiol.,* 87, 507, 1930.

58. **Shvartz, E. and Meyerstein, N.,** Tilt tolerance of young men and young women, *Aerospace Med.,* 41, 253, 1970.

59. **Schvartz E. and Meyerstein, N.,** Effect of heat and natural acclimation to heat on tilt tolerance of men and women, *J. Appl. Physiol.,* 28, 428, 1970.

60. **Greenleaf, J. E., Brock, P. J., Sciaraffa, D., and Elizondo, R.,** Effects of exercise-heat acclimation on fluid, electrolyte, and endocrine responses to tilt and Gz acceleration in women and men, *Aviat. Space Environ. Med.,* 56, 683, 1985.

61. **Frey, M. A. B., Mathes, K. L., and Hoffler, G. W.,** Cardiovascular responses of women to lower body negative pressure, *Aviat. Space Environ. Med.,* 57, 531, 1986.

62. **Raven, P. B., Rohm-Young, D., and Blomqvist, C. G.,** Physical fitness and cardiovascular response to lower body negative pressure, *J. Appl. Physiol.: Respirat. Environ. Exercise Physiol.,* 56 138, 1984.

63. **Frey, M. A. B. and Hoffler, G. W.,** Association of sex and age with responses to lower-body negative pressure, *J. Appl. Physiol.,* 65, 1752, 1988.

64. **Hordinsky, J. R., Gebbard, U., Wegmann, H. M., and Schaefer, G.,** Cardiovascular and biochemical response to simulated space flight entry, *Aviat. Space Environ. Med.,* 52, 16, 1981.

65. **Goldwater, D. J, and Sandler, H.,** Orthostatic and acceleration tolerance in 55 and 65 Y.O. men and women after weightlessness simulation, *Preprints of Annual Scientific Meeting,* Aerospace Medical Association, Washington, DC, 1981.

66. **Montgomery, L. D., Kirk, P. J., Payne, P. A., Gerber, R. L., Newton, S. D., and Williams, B. A.,** Cardiovascular responses of men and women to lower body negative pressure, *Aviat. Space Environ. Med.,* 48, 138, 1977.

67. **Sandler, H. and Winter, D.,** Physiologic Response of Women to simulated weightlessness, National Aeronautics and Space Administration, Washington, D.C., NASA SP-430, 1978.

68. **Frey, M. A. B., Mathes, K. L., and Hoffler, G. W.,** Aerobic fitness in women and responses to lower body negative pressure, *Aviat. Space Environ. Med.,* 58, 1149, 1987.

69. **Hudson, D. L., Smith, M. L., and Raven, P. B.,** Physical fitness and hemodynamic response of women to lower body negative pressure, *Med. Sci. Sports Exerc.,* 4, 375, 1987.

70. **Torikoshi, S., Yokozawa, K., Inazawa, M., Itoh, K., and Fukase, Y.,** Effects of lean body mass and aerobic power on LBNP tolerance in women, *Physiologist,* 30 (Suppl.), S75, 1987.

71. **Sandler, H.,** Cardiovascular Effects of Weightlessness and Ground-Based Simulation, NASA Technical Report 88314, National Aeronautics and Space Administration, Washington, D.C., 1988.

72. **Montgomery, L., Goldwater, D. J., Rositano, S. A., and Sandler, H.,** Peripheral blood flow response of women (ages 35 to 45 years) to lower-body negative pressure as a consequence of nine days bed rest, *Preprints of Annual Scientific Meeting,* Aerospace Medical Association, Washington, DC, 1978, 144.

73. **Goldwater, D., Sandler, H., and Montgomery, L.,** Exercise capacity, hematology, and body composition of females during bedrest shuttle flight simulation, *Preprints of Annual Scientific Meeting,* Aerospace Medical Association, Washington, DC, 1978.

74. **Polese, A. D., Goldwater, D., and Rositano, S. A.,** Effect of bedrest and lower body negative pressure (LBNP) on left ventricular systolic time intervals, Reports of Annual Scientific Meeting, Aerospace Medical Association, Washington, DC, May 14-17, 1979.

75. **Gillingham, K. K., Schade, C. M., Jackson, W. G., and Gilstrap, L. C.,** Women's $+G_z$ tolerance, *Aviat. Space Environ. Med.,* 57, 745, 1986.

76. **Greenleaf, J. E, Brock, P. J., and Sciaraffa, D.,** Effect of physical training in cool and hot environments on $+G_z$ acceleration tolerance in women, *Aviat. Space Environ. Med.,* 56, 9, 1985.

77. **Berry, J. J, Montgomery, L. D., Goldwater, D., Bagian, J., and Sandler, H.,** Hemodynamic responses of women 46 to 55 years to $+G_z$ acceleration before and after bed rest, *Preprints of Annual Scientific Meeting,* Aerospace Medical Association, Washington, DC, 1980.

78. **Keil, L. C. and Ellis, S.,** Plasma vasopressin and renin activity in women exposed to bed rest and $+G_z$ acceleration, *J. Appl. Physiol.,* 40, 911, 1976.

79. **Greenleaf, J. E., Stennett, H. O., Davis, G. L., Kalias J., and Bernaeur, E. M.,** Fluid and electrolyte shifts in women during $+G_z$ acceleration after 15 days' bed rest, *J. Appl. Physiol.,* 42, 67, 1976.

80. **Sandler, H., Goldwater, D. J., and Rositano, S. A.,** Physiologic responses of female subjects during bed rest shuttle flight simulation, *Preprints of Annual Scientific Meeting,* Aerospace Medical Association, Washington, DC, 1978.

Chapter 6

ORTHOSTATIC HYPOTENSION

Mahendr S. Kochar

TABLE OF CONTENTS

I. INTRODUCTION

Upon assumption of the upright posture from a supine position, the blood can be expected to gravitate into the lower extremities, reducing the venous return to the heart and cardiac output, leading to a reduction in cerebral perfusion. In a healthy individual, this is prevented by complex regulatory mechanisms described in the previous chapters of this book. These mechanisms counteract the gravitational forces on the blood and maintain systemic arterial pressure and cerebral perfusion.

In short, the sudden transient fall in blood pressure upon assumption of the upright posture is perceived by the baroreceptors located in the aortic arch and carotid bifurcations. This information is carried to the vasomotor center in the brainstem by the afferent sympathetic nerves. This causes the vasomotor center to send out impulses via efferent sympathetic nerves to the heart, peripheral blood vessels, adrenals, and kidneys, leading to arteriolar and venous construction, increased heart rate, increased plasma catecholamines, and activation of the renin-angiotensin-aldosterone system. In addition, vasopressin secretion from the hypothalamus and posterior pituitary is stimulated and the secretion of atriopeptin (atrial naturetic factor [ANF]) is diminished. The net result of these reflex changes is an acceleration in the heart rate of 5 to 25 bpm and either no change in the systemic arterial blood pressure or a rise in mean systemic arterial blood pressure of 5 to 10 mmHg.

II. DEFINITION AND CLASSIFICATION OF ORTHOSTATIC HYPOTENSION

There is no firm agreement in the literature on the definition of orthostatic hypotension. McDowell defines postural hypotension as a fall in systolic blood pressure of 30 mmHg or to less than 80 mmHg upon standing.[1] Bannister regards the presence of orthostatic symptoms to be more relevant than the numerical fall in blood pressure.[2] An orthostatic drop in either systolic or diastolic blood pressure of >20 mmHg is regarded as significant by most authors.

In my judgment, it is not only important to define orthostatic hypotension, but also to classify it for monitoring the patient's condition and documenting improvement with treatment. Based on the symptoms accompanying the orthostatic drop in the blood pressure, I have devised afunctional classification for these patients, which is shown in Table 1.

III. CAUSES

A. ACUTE ORTHOSTATIC HYPOTENSION

Acute orthostatic hypotension in almost all instances results from a reduction in circulatory blood volume as a result of fluid depletion or blood loss. The common causes of fluid volume depletion are listed in Table 2. The treatment of acute orthostatic hypotension comprises measures to stop further volume depletion and administration of appropriate fluids to replace the losses.

In rare instances, acute autonomic neuropathy, a variant of acute idiopathic polyneuritis (Landry-Guillain-Barré syndrome) can cause orthostatic hypotension accompanied with other manifestations of sympathetic and parasympathetic nervous system dysfunction.[3,4] Recovery occurs within a few months, possibly hastened by prednisone therapy.

B. CHRONIC ORTHOSTATIC HYPOTENSION

Causes of chronic orthostatic hypotension are listed in Table 3. Antihypertensives and psychotropic drugs are probably the most common cause of chronic orthostatic hypotension. In most cases, the postural hypotension is mild and well tolerated. If postural hypotension is class 2 or more, a reduction in dose or change of the drug is recommended.

TABLE 1
Functional Classification of Orthostatic Hypotension

Class 1. Asymptomatic postural hypotension (fall in either systolic or diastolic BP ≥20 mmHg).

Class 2. Lightheadedness (dizziness, giddiness) associated with postural hypotension but no history of syncope.

Class 3. History of syncope (fainting) accompanied with postural hypotension.

Class 4. Incapacitated because of severe dizziness or frequent syncope due to documented postural hypotension.

TABLE 2
Causes of Fluid Volume Depletion

Sudden blood loss
Gastrointestinal losses
  Vomiting
  Gastric or small-bowel drainage
  Diarrhea
  Bowel fistulas (colostomy, ileostomy, etc.)
Renal losses
  Chronic renal failure
  Diuretic phase of acute tubular necrosis
  Postobstructive nephropathy
  Nephrotic syndrome
  Adrenal insufficiency
  Osmotic diuresis (diabetic glycosuria)
  Diuretics
Skin losses
  Sweating
  Burning
Paracentesis

Orthostatic hypotension due to diabetes mellitus is almost always associated with peripheral neuropathy and other manifestations of autonomic dysfunction.[5] Orthostatic hypotension is not a common clinical feature of diabetes insipidus as long as the patient has free access to water to keep fluid intake sufficiently high to prevent dehydration. Orthostatic hypotension is a common clinical feature of adrenal insufficiency, which should be excluded in all patients manifesting chronic orthostatic hypotension. Although orthostatic hypotension may be observed in patients with pheochromocytoma secreting predominately epinephrine (adrenaline) and/or dopamine, it is the episodic hypertension accompanied by sweating, pallor, and headache that constitutes the common presenting manifestations.

Almost any disease of the spinal cord can be accompanied by orthostatic hypotension. Many conditions usually manifesting as peripheral neuropathies can have accompanying autonomic neuropathy and orthostatic hypotension. These include amyloidosis, Guillain-Barré syndrome,[4] diabetes,[5] bronchogenic carcinoma,[6] rheumatoid arthritis,[7] pernicious anemia,[8] and uremia.[9] Parkinsonism is sometimes associated with mild postural hypotension that is made worse by levodopa. Shy-Drager syndrome is characterized by orthostatic hypotension and widespread neurological lesions of extrapyramidal, pyramidal, and cerebellar systems as part of multiple system atrophy, but the peripheral nerves are intact.[10] In addition to severe sympathetic dysfunction, parasympathetic dysfunction is also present.[11] In patients with idiopathic orthostatic hypotension, there is no involvement of the central nervous system or peripheral nerves, but there are other manifestations of autonomic neuropathy.[12] Ziegler et al. have reported that, in general, patients with Shy-Drager syndrome have normal resting plasma norepinephrine levels,

TABLE 3
Causes of Chronic Orthostatic Hypotension

1. Drugs:
   Antihypertensives: Diuretics, sympathetic inhibitors,
      vasodilators, ACE inhibitors, calcium blockers, nitrates
   Psychotropics: antidepressants, phenothiazines, sedatives
   Anti-Parkinsonians: levodopa, bromocriptine
2. Endocrine-metabolic disorders:
   Diabetes mellitus
   Diabetes insipidus
   Primary and secondary adrenal insufficiency
   Primary hypoaldosteronism
   Hyporeninemic hypoaldosteronism
   Pheochromocytoma
3. Neurogenic disorders:
   Multiple sclerosis
   Amyotropic lateral sclerosis (motor neuron disease)
   Syringomyelia
   Tabes dorsalis
   Peripheral neuropathies, especially due to amyloidosis
   Spinal-cord section or myelopathy
   Lumboscral sympathectomy
   Primary dysfunction of the afferent limb
   Parkinsonism
   Shy-Drager syndrome
   Idiopathic orthostatic hypotension
4. Mitral valve prolapse and cardiac disorders causing reduction
      in cardiac output.
5. Malnutrition, cachexia, chronic bedrest, large varicose veins,
      aging, alcohol, fever, pregnancy.
6. Normal variant in tall asthenic individuals and the elderly.
7. Miscellaneous disorders:
   Idiopathic sympathotonic hypotension
   Hyperbradykininism

while those with idiopathic orthostatic hypotension have depressed levels at rest; however, enough exceptions are seen that the norepinephrine concentration should not be used to make the diagnosis. Both groups fail to increase their circulating levels of the neurotransmitter during standing and exercise. Thus, it appears that those with Shy-Drager syndrome have an intact peripheral sympathetic nervous system but are unable to activate it, while those with idiopathic orthostatic hypotension have true insufficiency of the peripheral autonomic nervous system.[13]

Weight loss unrelated to fluid loss, particularly in patients with cancer or other chronic debilitating diseases, often results in a fall in blood pressure, including orthostatic hypotension. The mechanism of the fall in blood pressure is probably the opposite of the mechanism that causes a rise in blood pressure upon weight gain and remains ill understood.[14] Idiopathic sympathotonic hypotension is a poorly defined condition manifested by sinus tachycardia and hypotension upon standing. Sympathetic nerves are intact but there seems to be impaired α-receptor activation upon standing.[15] Streeten and associates have observed several patients who exhibited syncopal attacks associated with orthostatic hypotension and severe tachycardia. When recumbent, these patients commonly had flushing of the face, neck, and anterior chest. The investigators found the plasma concentration of bradykinin to be above normal, but plasma kallikreinogen was normal. They postulated that these patients had deficient kinin-inactivating enzymes, causing hyperbradykininemia, leading to arteriolar and venular dilatation causing postural hypotension. The sinus tachycardia is due to reflex sympathetic activation, as the sympathetic nervous system is intact.[16] A primary dysfunction of the afferent limb of the arterial baroreceptor reflex system has been reported to cause supine hypertension and orthostatic hypotension.[17] A postprandial blood pressure decrease in well elderly persons is well documented, but in some less robust elderly, it may be a factor in syncope and fall.[18]

IV. EVALUATION

The purpose of the evaluation of a patient with chronic orthostatic hypotension is to seek answers to the following questions:

1. Does the patient have orthostatic hypotension?
2. What is the severity of the orthostatic hypotension as determined by the functional classification stated previously?
3. Are there other manifestations of autonomic neuropathy?
4. What is the etiology of the orthostatic hypotension?
5. What other organ systems are involved?
6. Are there conditions present that would contraindicate certain drugs that could be necessary for the treatment of hypotension?

A. HISTORY

The history should elicit information about the patient's symptoms, particularly orthostatic symptoms of dizziness, vertigo, lightheadedness or giddiness. Patients should be specifically asked if they have experienced fainting or blackout spells. The question should be asked about other manifestations of autonomic insufficiency such as impotence, bladder dysfunction manifesting as retention or incontinence of urine, bowel dysfunction manifesting as diarrhea or constipation, dysphagia, impaired sweating, and symptoms of heart failure such as shortness of breath and leg edema. Specific inquiry should be made regarding a history of diabetes, cancer, alcoholism, renal disease, and peptic ulcer. Patients should be asked to provide a list of all the medications they are taking or, better still, actually show the medication bottles. A history of sympathectomy should be elicited. The physician should inquire about a family history of disorders associated with orthostatic hypotension.

B. PHYSICAL EXAMINATION

The physical examination should document blood pressure and heart rate in the supine, sitting, and standing positions in both arms, noting the occurrence of symptoms such as dizziness or fainting upon getting up from a recumbent posture. The standing blood pressure should be recorded immediately, 1 min and 3 min after standing, and after 5 min of ambulation. A complete physical examination should be performed looking for physical signs generally associated with conditions known to be accompanied by postural hypotension, as listed in Table 3. Other signs of autonomic insufficiency, such as poor sphincter tone or bladder distension, and anhydrosis in selective areas, such as the armpits, should be noted. A complete neurological examination is a must in an evaluation of orthostatic hypotension.

C. LABORATORY TESTS

Urinalysis — Proteinuria may indicate renal disease due to amyloidosis or diabetes.

A complete blood count should be performed to exclude anemia, as severe anemia can cause dizziness with or without postural hypotension. Macrocytic anemia may suggest folic acid deficiency due to alcoholism or vitamin B_{12} deficiency due to pernicious anemia; both conditions as previously stated, can cause orthostatic hypotension.

The blood chemistry should include measurements of blood urea nitrogen (BUN) and creatinine to rule out renal disease, fasting blood sugar for diagnosis of diabetes mellitus, and serum potassium for adrenal insufficiency, in which instance it is almost always high. If adrenal insufficiency is suspected, it should be confirmed by demonstrating a high plasma renin activity and a low plasma aldosterone after 2 h of ambulation. The cosyntropin (synthetic ACTH) stimulation test helps to differentiate primary from secondary adrenal insufficiency. Plasma catecholamines upon awakening in the morning and after a half hour of upright posture or ambulation, as stated previously, can in some cases help to differentiate between the Shy-Drager syndrome and idiopathic orthostatic hypotension. Glycosylated hemoglobin and a 2-h postprandial serum glucose test can help to diagnose early diabetes mellitus, but orthostatic hypotension as a manifestation of diabetes occurs only in advanced cases and, therefore, the glucose tolerance test (GTT), even if abnormal, is seldom helpful in determining the etiology of orthostatic hypotension.

A chest X-ray should be performed to rule out bronchogenic carcinoma, because orthostatic hypotension as a manifestation of autonomic neuropathy with or without peripheral neuropathy has been described.[6]

Electromyogram (EMG) and nerve conduction studies are performed for establishing the state of peripheral nerve functions. Tests for autonomic insufficiency are listed in Table 4. These are necessary only if the symptoms and signs of autonomic dysfunction are subtle or if one wishes to localize the site of the lesion in the autonomic nervous system.[19] Serum and urine protein electrophoresis help to establish the diagnosis of amyloidosis. A blood volume measurement is done if there is a question of chronic volume depletion.

V. THERAPY

Reversible causes must be identified and corrected first and foremost. Volume depletion, hypotension-inducing drugs, and adrenal insufficiency are some of the reversible causes. Strict control of blood sugar in diabetics can often slow the progression of neuropathy.

For symptomatic treatment of orthostatic hypotension, the following stepped care approach is effective in most patients (Table 5).

A. GENERAL MEASURES

Step 1 treatment of orthostatic hypotension is comprised of general nonpharmacological measures to increase the blood pressure. These include weight gain if nonobese, additional salt intake including salt tablets, and plenty of fluids. The patient should be advised to sleep with the head of the bed elevated to the maximal tolerated angle. This encourages local autoregulation of blood flow by cerebral vessels, chronically adapting them to low perfusion pressures, and avoids the sudden pooling of blood upon arising in the morning. The bed tilt also stimulates the renin-angiotensin-aldosterone system and vasopressin secretion, causing vasoconstriction and sodium and water retention to expand the vascular volume. The patient should be advised to have a urine bottle or bedpan available next to the bed to avoid having to get out of bed at night to urinate. Patients should be further advised to sit on the edge of the bed upon arising before standing up and should postpone activities such as shaving until they have been up for a while. Eating smaller meals more frequently with coffee can help avoid postprandial orthostatic hypotension.[20] Commercially available elasticized antigravity garments, such as elastic stockings, support leotards, and G suits, are extremely helpful in many patients.[21] For class 1 orthostatic hypotension, in some patients knee-length elastic stockings may prove sufficient in minimizing the orthostatic drop in blood pressure. Elasticized garments extending from the metatarsal to the costal margin allow a gradient of counterpressure to be applied, with maximal pressure at the ankles and slight counterpressure at the top. The garment, which can be made with an open crotch to facilitate urination and zippers on both sides to ease wearing, should be donned while the patient is still recumbent. In patients who are incapacitated, a mild degree of exercise under supervision is highly desirable to activate the muscle pump of the legs. These general nonpharmacological measures can raise the mean blood pressure by as much as 20 mmHg and

TABLE 4
Tests of Autonomic Function

Purpose of the Test	Procedure	Normal Response
1. Integrity of baroreceptor circuit	Tilt-Table test	Rise in HR, little change in BP
	Valsalva Maneuver	Four-phase response with overshoot of BP
	Graded neck suction	2 — 6 ms of RR interval prolongation per mmHg of applied suction
a. Response to pressor stimuli	I.V. bolus of 25 or 50 μg phenylephrine	Rise in BP, reflex fall in HR
b. Response to depressor stimuli	Amyl nitrate inhalation or	
	I. V. infusion of nitroprusside	Fall in BP, reflex rise in HR
2. Vasomotor center responsiveness	Hyperventilation	Reduction of BP
3. To test efferent sympathetic fibers:		
a. Cold pressor test	Immersion of both hands in ice-cold (4°C) water for 1 min	Rise in BP
b. Mental arithmetic	Serial subtraction of 7 from 100 while the patient's concentration is disturbed	Rise in BP
c. Sweat test	Exposure to hot environment	Sweating
4. To test extraadrenal stores of norepinephrine	I.V. infusion of tyramine	Rise in BP
5. Test for low-pressure receptors (cardiopulmonary receptors)	Lower body negative pressure	Rise in BP

Note: Continuous monitoring of intraarterial blood pressure (BP) and electrocardiogram for heart rate (HR) during these tests is highly desirable.

TABLE 5
Stepped-Care Treatment of Orthostatic Hypotension

Step 1. Salt, weight gain, bed tilt, antigravity garments, exercise
Step 2. Fluodrocortisone/vasopressin
Step 3. Vasoconstrictors with or without beta blockers
Step 4. Indomethacin or other NSAIDS
Step 5. Atrial tachypacing

TABLE 6
Drugs Used in Treatment of Orthostatic Hypotension

Fluid retainers
  Fluodrocortisone
  Vasopressin and analogs: lypressin, desmopressin
Vasoconstrictors
  Ephedrine
  Methylphenidate
  Phenylephrine
  Midodrine
  Tyramine with an MAO inhibitor
  Dihydroergotamine
  Metoclopramide
Beta blockers
  Propranolol and other nonselective beta blockers
  Pindolol and other beta blockers with ISA
Nonsteroidal antiinflammatory drugs (NSAIDs)
  Indomethacin
  Ibuprofen and other NSAIDs
Somatostain analogs
  SMS-201-995

should always be tried before administering drugs to raise the blood pressure. In addition, patients should be advised to do their chores 1 to 2 h prior to the next meal, when the blood pressure is higher, and rest for 30 to 60 min after eating, when the blood pressure is expected to be lower.

B. FIRST-LINE DRUGS

The drugs useful in the treatment of orthostatic hypotension are listed in Table 6. When the step 1 measures prove insufficient, fluodrocortisone (9-α-fluodrohydrocortisone) is the drug of choice. In low doses, 0.1 to 0.2 mg/d, the drug increases the sensitivity of blood vessels to intrinsic vasoconstrictors, such as noradrenaline and angiotensin II.[22] In higher doses, it works through the distal tubule of the kidney, enhancing sodium absorption, which leads to plasma volume expansion and, consequently, a rise in blood pressure. In doses higher than 0.4 mg/d, it can lead to congestive heart failure, peripheral edema, recumbent hypertension, and hypokalemia. Most class 2 patients can be made symptom free with a combination of step-1 and step-2 measures. Antidiuretic hormone (arginine vasopressin [AVP]) levels in many patients with orthostatic hypotension are low, and vasopressin has been shown to play an important role in preventing or minimizing orthostatic hypotension in diabetic patients. Daily injections of ADH or nasal inhalations of an AVP analog such as lysine vasopressin (lypressin, Diapid) or desmopressin (DDAVP) can increase the blood pressure in a subset of these patients and should be tried if fluodrocortisone does not raise the blood pressure.[23-26]

C. VASOCONSTRICTING DRUGS

When the above two steps are insufficient to overcome orthostatic hypotension and symptoms persist, serious consideration should be given to the addition of one or more vasoconstricting drugs. These include ephedrine in dose of 25 mg orally four to six times a day, methylphenidate 20 mg twice a day, phenylephrine 10 to 20 mg three or four times daily orally or by a nasal spray, midodrine[27] in a dose of 2.5 to 10 mg three times daily, and dihydroergotamine[28] in a dose of 2 to 3 mg three times daily. Midodrine is an α-adrenergic agonist and dihydroergotamine is an α antagonist that has a considerable partial agonist effect. Dihydroergotamine has been postulated to partly act by stimulating local synthesis of vasoconstrictor prostaglandins.[29] Monoamine oxidase inhibitors, such as tranylcypromine in a dose of 10 mg two or three times daily, have been used in combination with tyramine, 2 to 6 mg three times daily,[30] but this therapy should be used with extreme caution and under expert supervision, as severe supine hypertension can result.[31] β antagonists have been used successfully in hyperbradykininemia.[16] The β antagonist, pindolol, which has a partial agonistic activity, is reported to be useful in some patients, especially in those with a very low supine norepinephrine concentration.[32] It may be particularly useful in patients with supine hypertension and orthostatic hypotension. Caution is advised as heart failure may develop.[33] For the control of nocturnal supine hypertension, a dose of 10 to 20 mg nifedipine or 25 mg hydralazine can be administered if the systolic blood pressure is above 180 mmHg or the diastolic blood pressure is above 105 mmHg. The dopamine antagonist, metoclopramide, 10 mg three times daily, has been also used successfully in a few patients.[34] These drugs should be tried one after the other until one of them raises the standing blood pressure sufficiently and is tolerated well by the patient. More than two vasoconstrictors should not be administered simultaneously, as they can produce an unacceptably high supine blood pressure.

D. INDOMETHACIN

Indomethacin, a prostaglandin synthetase inhibitor, has been used successfully in the treatment of several patients with severe chronic orthostatic hypotension.[35,36] When the first three steps prove inadequate or the drugs are poorly tolerated by the patient, and in the absence of a history suggesting peptic ulcer or gastrointestinal bleeding, serious consideration should be given to using indomethacin in a dose of 75 to 150 mg per day, always administered on a full stomach to avoid gastrointestinal discomfort. Because of the high incidence of gastrointestinal side effects, including bleeding, indomethacin should be used with great caution. If the orthostatic hypotension responds to indomethacin but the drug is poorly tolerated, other nonsteroidal antiinflammatory agents, such as meclofenamate, ibuprofen, or naproxen, may be tried one after the other until the desired effect is obtained without side effects.[37] Although the other nonsteroidal antiinflammatory agents are tolerated better than indomethacin, they do not raise the blood pressure as much as indomethacin.

E. ATRIAL TACHYPACING

Atrial tachypacing using a transvenous coronary-sinus atrial demand pacemaker set at a rate of 70 to 100 bpm is recommended as a last resort when other measures fail.[38,39] A temporary pacemaker should be first tried to make sure it is helpful prior to the insertion of a permanent pacemaker.

An injectable long-acting somatostatin analog, SMS-201-995, has been shown to raise postprandial blood pressure, but its place in the management of orthostatic hypotension has not yet been defined.[40]

The above-described therapeutic measures offer only symptomatic relief, and the underlying disease may continue to progress. In severe orthostatic hypotension, one may not be able to completely normalize the blood pressure, but often one can improve the quality of a patient's life by increasing the blood pressure sufficiently to allow the patient to ambulate. Eventually, despite intensive therapy, it might not be possible to raise the blood pressure sufficiently for the patient to ambulate. In the future, it may be possible to define specific abnormalities that can predict the response to the various drugs, and ways may be found to promote the regeneration of the autonomic nervous system.

VI. CONCLUSIONS

It is helpful for the purpose of diagnosis, treatment, and follow-up to classify orthostatic hypotension based on the severity of the symptoms. An earnest attempt should be made to identify and treat the reversible causes of orthostatic hypotension based on the severity of the symptoms. For symptomatic treatment, a stepped-care approach is recommended, starting with simple nonpharmacological measures. Fludrocortisone and vasoconstrictors should be added, one after the other as needed. Atrial tachypacing is reserved for patients whose response to more conservative measures is insufficient and in whom raising the heart rate with a temporary pacemaker is demonstrated to be helpful.

REFERENCES

1. **McDowell, F. H,** Orthostatic hypotension, in *L-Dopa and Parkinsonism*, Barbeau, A. and McDowell, F. H., Eds., F. A. Davis, Philadelphia, 1970, 263.
2. **Bannister, R.,** Chronic autonomic failure with postural hypotension, *Lancet,* ii, 404, 1979.
3. **Hopkins, A., Neville, B., and Bannister, R.,** Autonomic neuropathy of acute onset, *Lancet,* i, 769, 1974.
4. **Edmonds, M. E. and Sturrock, R. D.,** Autonomic neuropathy in the Guillain-Barre syndrome, *Br. Med. J.,* 2, 668, 1979.
5. **Ewing, D. J., Campbell, I. W., and Clark, B. F.,** The natural history of diabetic autonomic neuropathy, *Q. J. Med.,* 49, 95, 1980.
6. **Green, G. J., Breckenridge, A. M., and Wright, F. K.,** Severe orthostatic hypotension associated with carcinoma of the bronchus, *Postgrad. Med. J.,* 55, 426, 1979.
7. **Edmunds, M. E., Jones, T. C., Saunders, W. A., and Sturrock, R. D.,** Autonomic neuropathy in rheumatoid arthritis, *Br. Med. J.,* ii, 173, 1979.
8. **Kalbfleisch, J. M. and Woods, A. N.,** Orthostatic hypotension associated with pernicious anemia, *JAMA,* 182, 198, 1962.
9. **Campese, V. M., Procci, W. R., Levitan, D., Romoff, M. S., Goldstein, D. A., and Massry, S. G.,** Autonomic nervous system dysfunction and impotence in uremia, *Am. J. Nephrol.,* 2, 140, 1982.
10. **Shy, M. G. and Drager, G. A.,** A neurological syndrome associated with orthostatic hypotension: a clinical-pathologic approach, *Arch. Neurol.,* 2, 511, 1960.
11. **Khurana, R. K., Nelson, E., Azzarelli, B., and Garcia, J. H.,** Shy-Drager syndrome: diagnosis and treatment of cholinergic dysfunction, *Neurology,* 30, 805, 1980.
12. **Bradbury, S. and Eggleston, C.,** Postural hypotension: a report of three cases. *Am. Heart J.,* 1, 73, 1925.
13. **Ziegler, M. G., Lake, C. R., and Kopin, I. J.,** The sympathetic nervous system defect in primary orthostatic hypotension, *N. Engl. J. Med.,* 296, 293, 1977.
14. **Peterson, H. R., Rothschild, M., Weinberg, C. R., Fell, R. D., Mcleish, K. R., and Pfeifer, M. A.,** Body fat and the activity of the autonomic nervous system, *N. Engl. J. Med.,* 318, 1077, 1988.
15. **Polinsky, R. J., Kopin, I. J., Ebert, M. H., and Weise, V.,** Pharmacological distinction of different orthostatic hypotension syndromes; *Neurology,* 31, 1, 1981.
16. **Streeten, D. H. P., Kerr, L. P., Kerr, C. B., Prior, J. C., and Dalakos, T. G.,** Hyperbradykininism: a new orthostatic syndrome, *Lancet,* ii, 1048, 1972.
17. **Kochar, M. S., Ebert, T. J., and Kotrly, K. J.,** Primary dysfunction of the afferent limb of the arterial baroreceptor reflex system in a patient with severe supine hypertension and orthostatic hypotension, *J. Am. Coll. Cardiol.,* 4, 802, 1984.
18. **Peitzman, S. J. and Berger, S. R.,** Postprandial blood pressure decrease in well elderly persons, *Arch. Intern. Med.,* 149, 286, 1989.
19. **Ibrahim, M. M.,** Localization of lesion in patients with idiopathic orthostatic hypotension, *Br. Med. J.,* 37, 868, 1975.
20. **Onrot, J., Golberg, M. R., Biaggioni, I., Hollister, A. S., Kincaid, D., and Robertson, D.,** Hemodynamic and humoral effects of caffeine in autonomic failure and therapeutic implications for postprandial hypotension. *N. Engl. J. Med.,* 313, 549, 1985.

21. **Sheps, S. G.,** Use of an elastic garment in the treatment of orthostatic hypotension. *Cardiology, 61 (Suppl.)*, (1), 271, 1976.

22. **Schmid, P. G., Eckstein, J. W., and Abboud, F. M.,** Effect of 9-fluodrohydrocortisone on forearm vascular response to norepinephrine. *Circulation,* 34, 620, 1966.

23. **Kochar, M. S.,** Hemodynamic effects of lysine-vasopressin in orthostatic hypotension, *Am. J. Kidney Dis.,* 6, 49, 1985.

24. **Mathias, C. J., Fosbraey, P., Da Costa, D. F., Thornley, A., and Bannister, R.,** The effect of desmopressin on nocturnal polyuria, overnight weight loss and morning postural hypotension in patients with autonomic failure, *Br. Med. J.,* 293, 353, 1986.

25. **Zerbe, R. L., Henry, D. P., and Robertson, G. L.,** Vasopressin response to orthostatic hypotension: etiologic and clinical implications. *Am. J. Med.,* 74, 265, 1983.

26. **Saad, C. I., Ribeiro, A. B., Zanella, M. T., Mulinari, R. A., Gavras, I., and Gavras, H.,** The role of vasopressin in blood pressure maintenance in diabetic orthostatic hypotension, *Hypertension,* 11, (Suppl. I), I217, 1988.

27. **Kaufmann, H., Braunan, T., Krakoff, L., Yahr, M. D., and Mandeli, J.,** Treatment of orthostatic hypotension due to autonomic failure with a peripheral alpha-adrenergic agonist (midodrine), *Neurology,* 38, 951, 1988.

28. **Jennings, G., Esler, M., and Holmes, R.,** Treatment of orthostatic hypotension with dihydroergotamine, *Br. Med. J.,* ii, 307, 1979.

29. **Muller-Schweinitzer, E.,** Studies on peripheral mode of action of dihydroergotamine in human and canine veins, *Eur. J. Pharmacol.,* 27, 231, 1974.

30. **Nanda, R. N., Johnson, R. H., and Keogh, H. J.,** Treatment of neurogenic orthostatic hypotension with a monoamine oxidase inhibitor and tyramine, *Lancet,* ii, 1164, 1976.

31. **Davies, B., Bannister, R., and Sever, P.,** Pressor amines and monoamine oxidase inhibitors for treatment of postural hypotension in autonomic failure: limitations and hazards, *Lancet,* i, 172, 1978.

32. **Man in't Veld, A. J. and Schalekamp, M. A. D. H.,** Pindolol acts as beta receptor agonist in orthostatic hypotension: therapeutic implications, *Br. Med. J.,* 282, 929, 1981.

33. **Davies, B., Bannister, R., Mathais, C., and Sever, P.,** Pindolol in postural hypotension: the case for caution, *Lancet,* 2, 982, 1981.

34. **Kuchel, O., Bun, N. T., Gutkowska, J., and Genest, J.,** Treatment of severe orthostatic hypotension by metoclopramide, *Ann. Intern. Med.,* 93, 841, 1980.

35. **Kochar, M. S. and Itskovitz, H. D.,** Treatment of idiopathic orthostatic hypotension and Shy-Drager syndrome with indomethacin, *Lancet,* i, 1011, 1978.

36. **Abate, G., Polimeni, R. M., Cuccurullo, F., Puddu, P., and Lenzi, S.,** Effects of indomethacin on postural hypotension in Parkinsonism, *Br. Med. J.,* 2, 1466, 1979.

37. **Watt, S. J., Tooke, J. E., Perkins, C. M., and Lee, M.,** The treatment of idiopathic orthostatic hypotension: a combined fludrocortisone and flurbiprofen regime, *Q. J. Med.,* 50, 205, 1981.

38. **Moss, A. J., Glaser, W., and Topol, E.,** Atrial tachypacing in the treatment of a patient with primary orthostatic hypotension, *N. Engl. J. Med.,* 302, 1456, 1980.

39. **Kochar, M. S.,** Simultaneous treatment of Raynaud's phenomenon and orthostatic hypotension, *Am. J. Med.,* 75, 537, 1983.

40. **Hoeldtke, R. D., O'Dorisio, T. M., and Boden, G.,** Treatment of autonomic neuropathy with a somatostain analogue SMS-201-955, *Lancet,* 2, 602, 1986.

INDEX